W9-COU-024

WOMEN, WORK AND FAMILY IN THE BRITISH, CANADIAN AND NORWEGIAN OFFSHORE OILFIELDS

Also by Jane Lewis

LABOUR AND LOVE: Women's Experience of Home and Family, 1850–1940 (*editor*)
THE POLITICS OF MOTHERHOOD
WOMEN IN ENGLAND: Sexual Divisions and Social Change, 1870–1940
WOMEN'S WELFARE/WOMEN'S RIGHTS (*editor*)
WHAT PRICE COMMUNITY MEDICINE?

Also by Marilyn Porter

HOME, WORK AND CLASS CONSCIOUSNESS

Women, Work and Family in the British, Canadian and Norwegian Offshore Oilfields

Edited by
Jane Lewis
Lecturer in Social Administration, London School of Economics
Marilyn Porter
Department of Sociology, Memorial University of Newfoundland
and
Mark Shrimpton
Research Unit Director, Community Services Council of Newfoundland and Labrador

St. Martin's Press New York

First published in the United States of America in 1988

Printed in Hong Kong

ISBN 0–312–00528–8

Library of Congress Cataloging-in-Publication Data
Women, work, and family in the British, Canadian, and Norwegian offshore oilfields.
Proceedings of the International Conference on Women and Offshore Oil held at Memorial University of Newfoundland, Canada, in late 1985.
Bibliography: p.
Includes index.
1. Offshore oil industry—Social aspects—Congresses. 2. Women oil industry workers—Congresses. 3. Oil industry workers—Family relationships—Congresses. I. Lewis, Jane (Jane E.) II. Porter, Marilyn, 1942– . III. Shrimpton, Mark. IV. International Conference on Women and Offshore Oil (1985 : St. John's Nfld.)
HD9560.5.W66 1988 305.4′3665 86–31393
ISBN 0–312–00528–8

Contents

Preface

This volume has emerged from an International Conference on Women and Offshore Oil held at Memorial University of Newfoundland, Canada, late in 1985. The purpose of the conference was to provide a forum for social scientists from different countries to discuss the findings of a number of major empirical studies into women's experience of the offshore oil industry as workers and as wives and mothers. The results of five of these studies are brought together in this book: Robert Moore and Peter Wybrow's study for the British Equal Opportunities Commission on women's work in the North Sea; the Newfoundland Government Petroleum Directorate's work on the position of women workers in the Newfoundland offshore; the research programme of the Norwegian Work Institutes into offshore work and family life; the Institute of Medical Sociology and the Ross Clinic's study of 'intermittent husband syndrome' in Aberdeen; and the Community Resource Services Ltd's investigation of the family impacts of offshore commuting in Newfoundland. The work of drawing these cross-national experiences and discourses together has proved both demanding and stimulating.

It has been the aim of the editors and the contributors to locate the studies in the burgeoning literature on women's experience of work and the family and, in particular, to explore the complex relationship between the categories of 'work' and 'family' in the specific context of the offshore oil industry. It is our hope that the book will increase our grasp of these important issues as well as contributing to an international understanding of the social and economic impacts of the offshore oil industry.

Acknowledgements

We would like to thank the organisations that contributed financially to the International Conference on Women and Offshore Oil held at Memorial University of Newfoundland, Canada, late in 1985: the Department of the Secretary of State, Government of Canada; the Social Sciences and Humanities Research Council of Canada; the Institute of Social and Economic Research, Memorial University; Husky Oil Operations Ltd; Bow Valley Offshore; the Government of Newfoundland and Labrador; and Petro-Canada Exploration.

We would also like to thank the many contributors to the conference. In particular we are indebted to Arnlaug Leira of the Institute of Social Research in Oslo for her constant encouragement, valuable advice, and material on Norway that would otherwise have been hard for us to obtain.

JANE LEWIS
MARILYN PORTER
MARK SHRIMPTON

Notes on the Contributors

Dorothy Anger is currently writing a history of Micmac Indian settlement on the west Coast of Newfoundland. She recently prepared a report for the Province's Royal Commission on Employment and Unemployment. She was a member of the Social Research and Assessment Division of the Newfoundland and Labrador Petroleum Directorate.

Gary Cake is with the Inter-Governmental Affairs Secretariat of the Government of Newfoundland. He was formerly the manager of Social Research and Assessment with the Newfoundland Petroleum Directorate and he has conducted research on Newfoundland residents employed in the offshore oil industry and has recently completed a critical assessment of the literature on energy development and social problems.

David Clark is Director of the Scottish Marriage Guidance Council. He has written and published widely on the sociology of marriage, divorce and remarraige. His most recent book (with Jackie Burgoyne) is *Making a Go of It. A Study of Step-Families in Sheffield* (1984).

Richard Fuchs is Director of Research in the Department of Rural Agricultural and Northern Development (Government of Newfoundland) and was formerly Director of Social Research in the Newfoundland and Labrador Petroleum Directorate. He is the author of *Half a Loaf is Better than None: the Newfoundland rural development's movement's adaptation to the crisis of seasonal unemployment*. (1987)

Hanne Heen is a social anthropologist and a researcher at the Work Research Institutes in Oslo. Her recently published work includes *Social Integration and Safety on the Ekofisk Field* (1986).

Jane Lewis teaches in the Department of Social Administration at The London School of Economics and Political Science and is interested in both women's history and women and social policy.

She is the author of *Women in England, 1870–1940* (1984) and the editor of *Women's Welfare/Women's Rights* (1983).

Robert Moore is Professor of Sociology at the University of Aberdeen and his main research interests are in the fields of race relations and immigration policy. He is the author (with Peter Wybrow) of the recent Equal Opportunities Commission report, *Women in the North Sea Oil Industry* (1984).

Marilyn Porter teaches sociology at Memorial University of Newfoundland and has previously taught at the Universities of Bristol and Manchester. She has published in the areas of women, work and ideology and on women's lives and work in rural Newfoundland. She is the author of *Home, Work and Class Consciousness* (1983).

Mark Shrimpton is Director of the Research Unit at the Community Services Council in St John's Newfoundland and a principal of Community Resource Services (1984) Ltd. He is the author of a number of reports and papers on the socio-economic impacts of oil development on Atlantic Canada.

Jorun Solheim is a social anthropologist currently engaged in studies of the catering department and general development of platform organisation on Gullfaks A. She works at the Work Research Institutes, Oslo and has many recently published articles on work and family organisation, including 'Complexity and Communality on a North Sea Platform' (1983).

Keith Storey is an Associate Professor of Geography at Memorial University Newfoundland and a principal of Community Resource Services (1984) Ltd. He has been very involved in social and economic impact research and is the author of *Hibernia: Demographic Impacts* (1984).

Rex Taylor is a medical sociologist with the Medical Research Council's Medical Sociology Unit in Glasgow. He has taught sociology in North America and Britain and is a member of the Economic and Social Research Council's social affairs committee. His current research interests lie in the field of social gerontology.

Peter Wybrow is a Senior Lecturer in Social Administration at Teeside Polytechnic, and was formerly a researcher at the University of Aberdeen.

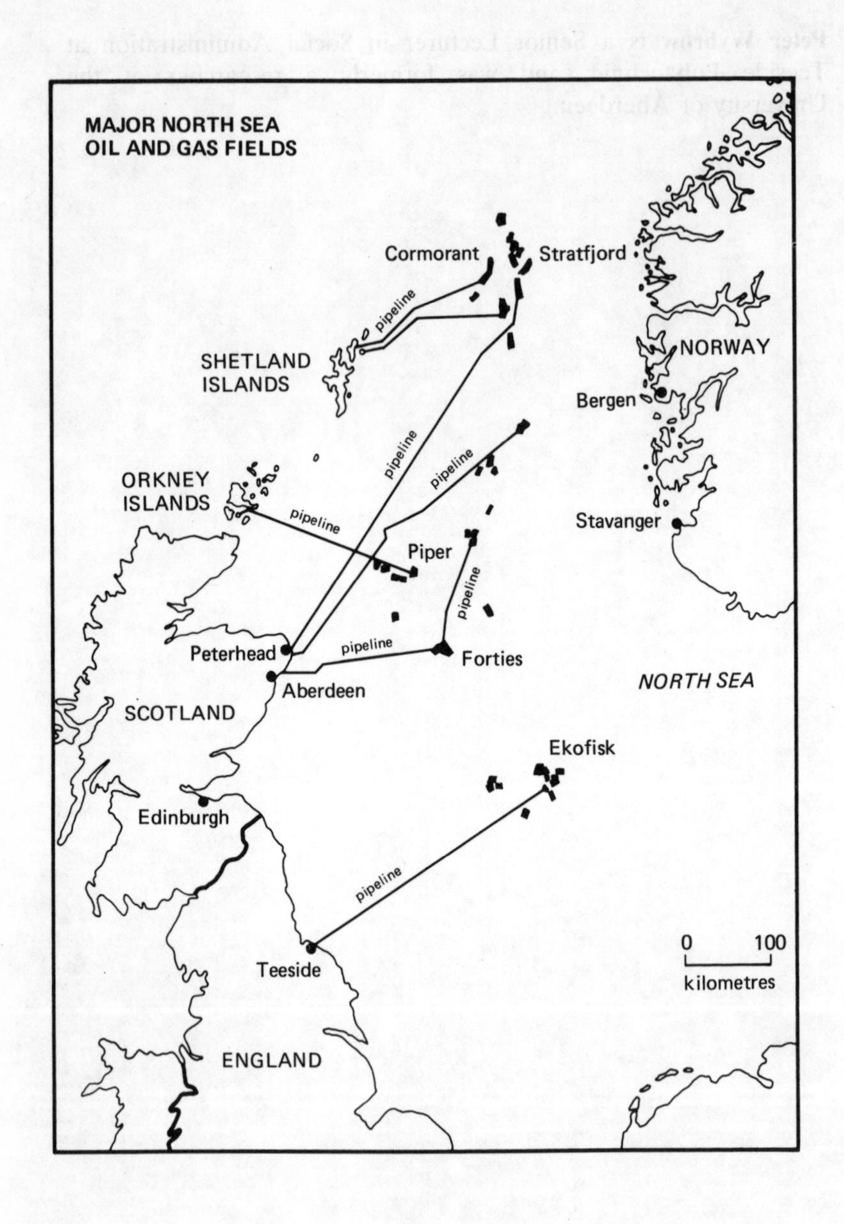
MAJOR NORTH SEA
OIL AND GAS FIELDS
Cormorant
Stratfjord
pipeline
SHETLAND
ISLANDS
NORWAY
Bergen
pipeline
pipeline
ORKNEY
ISLANDS
pipeline
Stavanger
Piper
pipeline
Peterhead
pipeline
Forties
Aberdeen
NORTH SEA
SCOTLAND
Ekofisk
Edinburgh
pipeline
0
100
kilometres
Teeside
ENGLAND

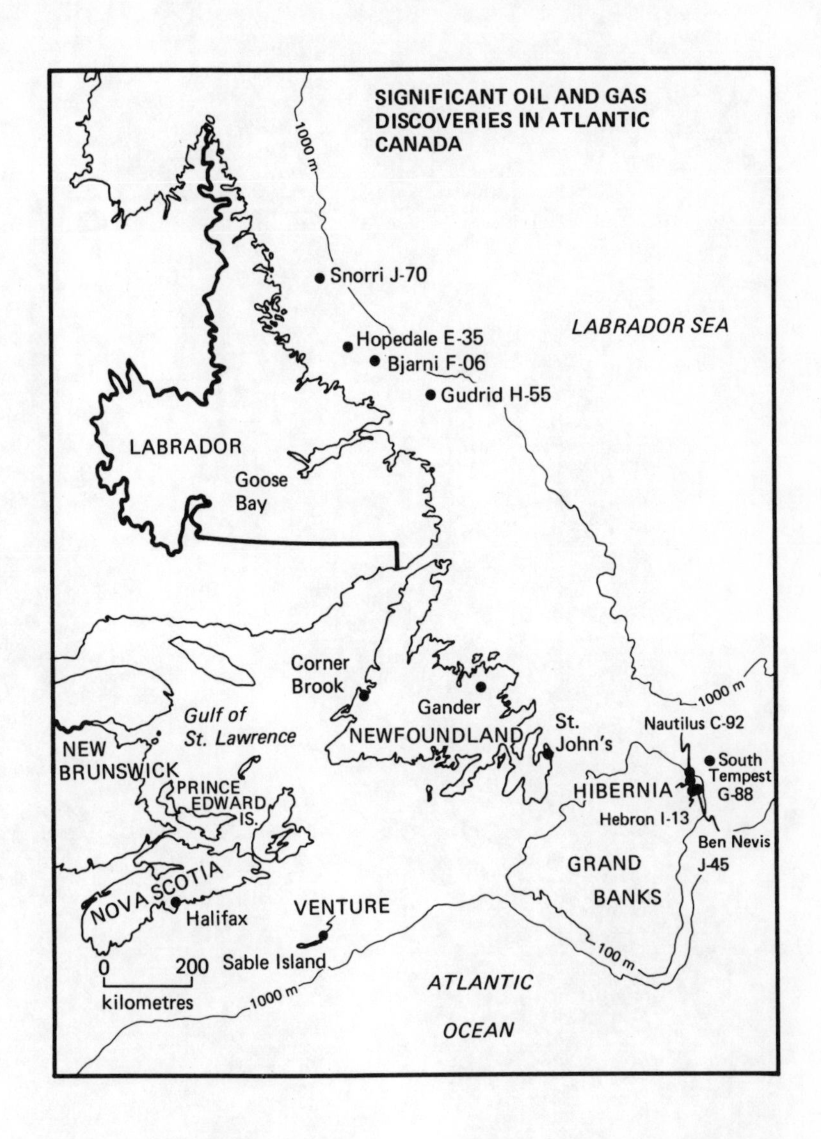
SIGNIFICANT OIL AND GAS DISCOVERIES IN ATLANTIC CANADA
1000 m
Snorri J-70
LABRADOR SEA
Hopedale E-35
Bjarni F-06
Gudrid H-55
LABRADOR
Goose Bay
Corner Brook
Gander
Gulf of St. Lawrence
NEWFOUNDLAND
St. John's
1000 m
Nautilus C-92
South Tempest G-88
HIBERNIA
Hebron I-13
Ben Nevis J-45
NEW BRUNSWICK
PRINCE EDWARD IS.
NOVA SCOTIA
Halifax
VENTURE
Sable Island
GRAND BANKS
100 m
0 200
kilometres
1000 m
ATLANTIC
OCEAN

General Introduction*

Editors of collections of essays such as this always face a problem in organizing the material so as to present the reader with manageable and coherent entities while, at the same time, preserving the overall logic that gave rise to the collection in the first place. In this case, by concentrating on the offshore oil industry[1] and then dividing the papers into those that focus on work and those that focus on the family, we stub our toes on two major problems, one of which is the question of treating the industry as if it were unique in all respects; the other of reiterating the conventional separation between 'family' and 'work'.

The oil industry as a whole undoubtedly has some remarkable features. The social and economic significance of oil exploitation is enormous. The industry is a complex multinational affair, distinguished primarily by its capacity to extract large amounts of both surplus value and profit. Its power and influence are reflected in the effects that oil price changes have had on international and national economies, and in the pressures that have been brought to bear on states which act counter to the industry's interests.

Despite its international nature, very few social scientists have cooperated to produce cross-national studies of its workings and impacts. These, in fact, vary considerably from place to place with very different effects on the work force and on family life. In part, such variation is dictated by the needs of the industry, but in part it has been dictated by the work-force and governments of particular countries and regions. As will be seen, Norway has been the most successful in modifying the work culture and practices of the offshore oil industry, and this in turn suggests that change is possible. This should also serve to alert us to the fact that people are not merely the passive recipients of offshore oil development. Their response involves an active interpretation of, and negotiation between, the social contours of their world and the possible positive or negative consequences of the new industry.

A second major distinguishing characteristic of the oil industry is

---

* We would like to acknowledge the substantial help given to us by Arnlaug Leira in writing this introduction.

its 'cowboy' image, which is exacerbated in the case of exploratory activity and in an offshore setting. Developing out of the southern United States, the industry has a highly individualistic ethos wherein any prospector or rig worker with 'the right stuff' may hope to rise to be the multi-millionaire president of the corporation. It is this image that is carried to the world by Hollywood films (not for nothing was oil well fire killer 'Red' Adair portrayed by John Wayne) and the popular television series *Dallas*: John D. Rockefeller and his successors are in many ways the archetypal folk heroes of capitalism.

The industry is portrayed as a tough male world, where emotions are kept on a tight rein and personal life has no place. As these essays show, there is considerable antipathy to any involvement of women. In terms of employment, they may be excluded (from the drill floor or the offshore drilling rig as a whole, at one extreme, and from Canada's most prestigious petroleum club – the fate of Pat Carney, Canada's Minister of Energy – at the other), or restricted to marginal occupations. With respect to the wives and children of male workers, the phrase 'we hire workers, not families' is heard on both sides of the Atlantic.

However, there is a danger in regarding the oil industry as entirely 'special'. There has been a tendency for local communities, such as those of Atlantic Canada, to be overly influenced by the image of the industry and to assume that the environmental and socio-economic impacts on the area will be different in every respect from previous local experience (House, 1985). But, while the potential impact of the oil activity is increased if it occurs in an area unused to large-scale industrial enterprise, not all aspects of the industry are quite new. In the case of the offshore oil activity, for instance, rotational work schedules have been part of the working life of the offshore fishermen and seamen for many years.

Similarly, the kinds of experiences women have had in the industry, both of hiring practices and of male work-place culture, are essentially similar to those in other traditionally male-dominated industries. In any case, many offshore oil-related jobs are onshore, and include such mundane positions as secretary, caterer, and personnel manager. Many offshore jobs are only unusual in their setting, not in the work itself, a point made particularly clear by Anger, Cake and Fuchs in this volume. In presenting these papers, we are not perpetuating any illusions that the oil industry is intrinsically different from other similarly constituted industries.

Nevertheless, the offshore oil industry on the fringes of the North Atlantic does provide us with an interesting focus and a useful case study. Its impact on marginal areas such as Northern Scotland, Norway, Newfoundland and Nova Scotia is highly visible. While the companies are not eager to have their work places researched (House, 1985, pp. 247–49), by comparing the experience women have of the oil industry (often involving the same multinational firms) in different but comparable economic, social and cultural contexts, we can broaden our understanding both of what happens in that kind of industry and of the mechanisms of development in economically marginal areas.

The division of papers in this volume into a 'work' and a 'family' section raises thorny issues for those who see one of the chief contributions of feminist theory in this area as being the collapse of these artificial boundaries, which emanate from and are validated within an inherently patriarchal scheme of things. We now recognise that women's experience is indivisible, that their potential or actual position as wives and mothers critically affects their role as workers and that the gendered division of labour in the home is rooted in economic dependence, which is itself a function of women's position in the labour market. In this regard, it makes no sense to follow the arbitrary pattern of dividing 'home' from 'work'. However, more recently, feminists have begun exploring the way in which work *itself* is gendered over and above real or imaginary biological needs of 'family' commitments (Roberts *et al.*, 1985; Beechey, 1985), and it is in line with this perception that we have explored the offshore oil industry where the very structure of work is so clearly organised around gender.

Another reason for the organisation of this volume is that, while when we set out to chart women's experience of the offshore oil industry we recognised that women would be affected both as workers and as members of the families of male workers, we did not fully appreciate what different categories of women there would be. Certainly not all women workers offshore are young or without family responsibilities. Indeed, the Newfoundland evidence suggests that they are older than the male workers, and the Norwegian evidence shows more and more women using offshore work as an integral part of a family economic strategy. However, as domestic responsibilities are currently defined in all three societies, it is harder for women with young children to meet the particular demands of offshore work. The two parts of the book, then, talk about rather different groups of women, and empirical research to date has not addressed the ways in which women's family responsibilities constrain their entry to the industry as workers.

Women remain a tiny minority of offshore workers – 3.9 per cent in Norway, 1.7 per cent in Newfoundland and somewhere between 0.1 and 0.5 per cent in Britain. Work in the offshore oil industry clearly poses two problems for women: one is overcoming traditional discriminatory practices and cultures; the other is claiming a space in a relatively new and rapidly developing technology. We should note that women who subject themselves to such a dual struggle are not doing it for reasons of theoretical interest, but because both traditional male industries and technologically advanced ones can provide much higher levels of pay. To what extent the low participation rates of women are a product of employers' practices, women's characteristics as workers, or of the work culture is considered in the first section of this book, together with the extent to which the situation may be ameliorated by equal opportunities legislation. The second section of the book concentrates on the experiences of the women who are the wives of male offshore workers. The rotational employment patterns of offshore workers tend to make explicit issues regarding sex roles in marriage, for example, in respect to parenting and responsibility for household labour and decision-making. It is not our intention that the experiences of the two groups of women should be treated as analytically discrete. Indeed, the second part of this general introduction will outline the way in which feminist analysis has questioned the traditional way in which sociologists have treated home and family as separate entities and will signal some of the important linking themes we wish to explore in this volume. Furthermore, the real problems expressed by offshore oil workers and their families regarding making the transition between home and work must be understood in the context of those larger economic and social forces which structure women's position in society. Offshore oil work is far removed physically and culturally from the world of home and family. It therefore provides an extreme, but socially and economically significant, filter for observing first the efforts of women to enter it as workers and to reap some of the material rewards it offers, and second the nature of the relationship between male workers in the industry and their families.

THE OFFSHORE OIL INDUSTRY

In examining the implications of the offshore oil industry for women it is important to recognise that the industry is not a monolithic and undifferentiated entity, but consists of a very large number of different types of companies engaged in a wide range of activities.[2] Most important are the oil companies themselves, which usually explore for, produce, transport, refine and retail petroleum. The largest and most famous of these are the oil 'majors' (the 'seven sisters': Exxon, Shell, Mobil, BP, Texaco, Chevron and Gulf). These are multinational enterprises whose size is traditionally measured by comparison with the gross national product of significant industrial nations. All seven are among the ten largest corporations in the world and control over 40 per cent of Western refining and marketing. In addition, there are large numbers of 'independents' and state oil companies of various sizes. The independents may be international or national, and if the latter, may be developed and supported as instruments of state policy. In the case of state oil companies (such as Statoil of Norway, Pemex of Mexico, Petrobras of Brazil and Petro-Canada), there is direct ownership.

However, while it is the oil companies that usually own or have rights to offshore areas containing (or thought likely to contain) petroleum reserves, the majority of work in exploring for and developing them is contracted out to other companies. Some, like the firms which own and operate the floating drilling rigs, are themselves major multinationals. At the other extreme are myriad small companies which, for example, might only have a catering or maintenance contract with respect to a single drilling rig operating in a particular region.

The international nature of the oil industry is particularly marked in the case of offshore activity, where much of the equipment used is common to the North Sea and Atlantic Canada or indeed to Alaska, the Gulf of Mexico, Indonesia, Australia, China and the Red Sea. Often it is literally the same equipment. The floating drilling rigs move internationally from one 'oilpatch' to another, and other expensive specialised equipment, such as pipelaying vessels and floating heavy-lift cranes, is similarly peripatetic. The helicopters, which fly out the crews and some equipment to the rigs and product platforms, and the supply boats, which ship equipment and materials from an onshore supply base, are more likely to be operated by smaller national companies, but there is a similar mobility of hardware. Even the supply bases themselves, whether in Peterhead, Scotland; Stavanger, Norway; St John's, Newfoundland; or elsewhere in the world, look much the same, with lengths of drill pipe, silos of drilling mud and cement and pieces of specialised equipment, supplied and supported by mostly United States international oil industry service companies.

Just as much of the equipment is highly mobile and international, so are many of the more senior and skilled personnel. In the most extreme case, oil well 'blow-out' fire killing specialists will be on 24-hour call to fly anywhere in the world. Toolpushers, the highly-paid workers who control the actual drilling operation, will often commute intercontinentally, as may some or all of the rest of the crew of the rig. Onshore and offshore executives can expect to move regularly and at short notice, either to another oilpatch or to a major administrative centre. To help support such a highly mobile existence, often involving living in remote locations, different cultures and extreme climates, most oil towns have a petroleum club, an oil wives' club and (in the case of larger centres) an 'American' school for the worker and his family. In many ways such employment is like working in the military, moving from one relatively self-contained 'base' to another.

The degree to which this multinational industry will be integrated into the local economy and society, and *vice versa*, depends on such things as the type and amount of activity it is engaged in, the length of time it has been in the region, the nature of the local economy and society, and the role of the state. Such an integration is of particular significance in the present context since it may serve to dilute the culture of the industry, with spin-off effects on hiring and other practices which traditionally have been antipathetic to any direct or indirect involvement of women.[3]

Offshore oil activity is of three types: exploration, development and production. The first of these involves searching for oil and gas by the use of seismic surveys (to determine the sub-sea geology) and then drilling wells. The drilling is mostly done by 'jack-up' drilling barges or 'semi-submersible' rigs. The former are held above the waves by retractable lattice steel legs which rest on the seabed; they are used in shallower waters such as are found in the southern North Sea and off Nova Scotia in Canada. Semi-submersible rigs float on submerged pontoons, are held in position by anchors or very powerful 'thruster' propellors, and have been responsible for most of the exploration drilling off Scotland, Norway and Newfoundland.

Such rigs are very expensive to build and operate; the drilling of recent Newfoundland exploration wells, for instance, cost an average of nearly £25 million[4] each (Canada, 1985). While the rigs may belong to oil companies, they are more usually owned and operated by one of the specialist drilling companies, many of which are based in the southern United States. Oil companies contract them to undertake drilling, usually on a per well basis, thereby placing a considerable premium on speed of operation. The rigs operate twenty-four hours a day, with most of the crew of about eighty being split into two twelve-hour shifts. For every period spent working offshore (and the length of this period varies, but is most commonly seven, fourteen, twenty-one or twenty-eight days) a similar amount of paid shore leave is normal.

In the early days of exploration there may be very little local involvement, with virtually all equipment and personnel shipped in. This may change over time, however, if local companies and workers are trained to provide the services and labour the industry requires. This will be reinforced if the state is able to enforce any local preference requirements it may have attached to the drilling licence. Lastly, local involvement will rise as the scale of activity increases, as this justifies local provision of goods and services (whether by indigenous businesses or branches of international companies).

However, a major increase in local involvement can be expected if exploration leads to the discovery of a commercial oil or gas field. This will result in development activity, whereby production equipment is designed, built, put into place, and brought into operation. The major component of this equipment is normally a production platform, which consists of a steel or concrete tower (the 'jacket') which is permanently affixed to the seabed and supports a deck structure high above the sea. The 'topsides' on this deck may contain equipment to drill production wells (this may instead, or additionally, be done by floating or jack-up rigs), but the focus of operations is the processing, storage and transhipment of oil and/or gas. Production facilities vary greatly in size, but a large field may require several hundred offshore workers on a number of platforms.

Development of an oil or gas field may be extremely expensive (the development of the Norwegian–United Kingdom Statfjord field is estimated to have cost £1.8 billion (Mackay, 1984), while Mobil Oil Canada (1985) have estimated that developing Newfoundland's Hibernia oil field using a single concrete platform would cost £2.2 billion) and create large numbers of jobs; the size and nature of economic, including employment, impacts on the region will depend on the size of the project and how much work is done locally. Some site and fabrication activity has to take place locally, and some components such as a concrete production platform, cannot be towed long distances. Beyond this, the primary determinants of local involvement will be the local industrial and labour capacity, and the role of the state. The state may seek to enforce local preference regulations and/or to ensure a national share in ownership of the project.

The decision to proceed to development and production also has, in and of itself, significant implications for the integration of the industry with the local economy and society. In the first place, development represents a considerable long-term capital commitment on the part of the oil companies involved. An exploration programme can be terminated at short notice, with the armada of equipment and the personnel very rapidly disappearing over the horizon. This may happen (or be threatened) because of poor drilling results, better prospects or state incentives elsewhere, a global recession in exploration, or because the state is being 'unreasonable' in its taxation, employment, environmental protection or other policies. Clearly a major long-term commitment to production weakens the industry's negotiating position in this regard, and increases the likelihood that it will be responsive to local requirements and priorities. Furthermore, the fact that the production platforms, unlike the exploration rigs, cannot move off to operate in another jurisdiction increases the possibility of labour unions organising the work-force.

A second implication of a decision to develop a field is that it leads to relatively long-term employment, with large offshore fields remaining in production for twenty or more years. From an industry perspective, while labour transportation is a very small component of total costs, logistical and economic advantages lead to an increasing use of local labour. From the perspective of local workers, businesses and the state, investment in training and industrial infrastructure makes more sense for longer-term projects.

Thus, all other things being equal, there will be an increasing integration of the industry with the local economy and society as time passes, particularly if development and production occurs. (Exploration activity will usually continue parallel to development and production, and remains very vulnerable to economic and political forces.) In the case of the regions considered in this volume, exploration started at about the same time (1966 in Atlantic Canada, 1967 in the North Sea),[5] but poor initial drilling results and the complications icebergs posed for Newfoundland operations meant that drilling proceeded much more slowly in Atlantic Canada. By the end of 1984 nearly 1500 exploration and appraisal wells alone (that is wells drilled to discover and establish the size of fields, but excluding production wells) had been drilled in the North Sea off Scotland and Norway (Great Britain, Department of Energy, 1985), compared to only 211 off Atlantic Canada (Canada, 1985), with many of the latter resulting from a generous federal drilling incentive programme. While North Sea production started in 1975 with thirty-eight oil and gas fields in production in 1984, Atlantic Canada has not yet moved beyond exploration activity despite some major discoveries.[6]

While the level of local integration of the offshore oil industry is related to the type, scale and duration of activity, it is clearly also related to the characteristics of the region and policies of the state involved. Reference has already been made to the influence of the labour market and industrial capacity on local employment levels and business involvement, but the role of the state can also be of considerable significance. First, as has already been noted, the state may be successful in ensuring a greater local involvement by use of employment and business preference policies, and this may serve to dilute the antipathy to the involvement of women which the industry has traditionally exhibited. Secondly, the state may succeed in legislating industry practices, in such areas as safety, collective bargaining and equal opportunities, so as to bring them in line with standards deemed desirable onshore. With respect to both, there are marked differences between the three regions considered in this book.

The Norwegian government has always been willing to intervene directly in the offshore oil industry, and is the archetype of what Noreng (1980) has called the 'North Sea model'. This sees a rejection of both uncontrolled capitalism and complete nationalisation, preferring a system of highly regulated capitalist development combined with significant direct participation through a state oil company. House has summarised the arguments for such an approach thus:

> Ideally the North Sea model minimizes the excesses and exploitative features of free enterprise capitalism. It channels oil industry initiatives so as to promote national economic and social ends, such as employment creation and business development. The national state captures a major share of the economic rent and legislates to protect the environment, ensure safety of personnel and minimize the disruption of oil-related development. At the same time the competitiveness, flexibility, efficiency, technological progressiveness and entrepreneurial initiative of the private capitalist system (as compared to a monolithic state monoply) are preserved.
>
> (House, 1985, p.109)

Norway has continued to pursue these objectives, with Statoil, the national oil company founded in 1972, taking increasing control of production activities. While employment creation was not at first a high priority, given a tight labour market, there has been an increasing attempt to 'Norwegianise' the industry (Lind and Mackay, 1980). This has included reducing foreign involvement in the ownership of, and employment in, the industry, and trying to ensure that Norwegian work practices, including the comprehensive Norwegian policies with respect to the employment of women, are adopted offshore.

In contrast to this 'controlled' development, the British government has generally given the private sector a free rein to develop the offshore as rapidly as possible, with the goals of generating tax revenues and improving the national balance of payments. In 1976 the newly elected Labour government did establish the British National Oil Company, modelled on Statoil, but this was later privatised by the Conservative government of Margaret Thatcher. Despite high levels of regional unemployment no use has been made of local preference policies, although the highly industrialised nature of the British economy, allied to the promotional efforts of agencies such as the Offshore Supplies Office (Mackay, 1984), may have made this unnecessary. The regulation of safety and other practices offshore has often been lax (Carson, 1982), and 1975 equal opportunities legislation still has not been extended to the offshore.

The situation in Canada is more complex and likely less familiar to most readers. Canada is a federal state, and questions of federal and provincial jurisdiction are central to Canadian politics,[7] and made more complicated in the case of Newfoundland by the fact that it did not become a part of Canada until 1949. For instance, it was only in 1985 that an agreement was reached with respect to the issue of which level of government had jurisdiction over the continental shelf off Newfoundland's coast.

This was a matter of considerable contention because of the competing priorities of the federal and provincial governments. Newfoundland is best known for its fishery, and also has significant forestry and mining industries, but its economic base has long been very weak with little manufacturing of any kind. Fiscal transfers from the federal government account for 48 per cent of provincial government revenues, and Newfoundland has consistently had the highest unemployment rate (the official rate averages about 20 per cent, and the real rate is significantly higher, especially in rural areas), tax rates and cost of living of any Canadian province. Male and female labour force participation rates and levels of formal education are low compared to the rest of Canada, and there is a tradition of temporary and permanent out-migration to find employment. It is probably fair to say that unemployment is *the* social, economic and political 'fact of life' in Newfoundland.

In 1977 the provincial government sought to regulate the fledgling offshore oil industry by passing legislation which has been described as 'an application of the North Sea model to the particular circumstances of Newfoundland' (House, 1985, p.56). This established a government-owned Newfoundland and Labrador Petroleum Corporation and a tax regime which sought to maximise the province's share of the economic rent from any offshore petroleum produced. In addition, it placed major emphasis on employment creation, requiring oil companies to give preference to Newfoundland labour and companies, institute vocational training programmes, promote local research and development, and maximise the amount of oil and gas processing in the province. It also required the industry to undertake development at a pace and in a manner which would minimise negative social, economic and environmental impacts. The initial response of the oil industry was to withdraw the rigs, and no wells were drilled in 1977. Subsequently, however, they undertook drilling under both the competing provincial and federal terms.

The neighbouring province of Nova Scotia followed the interventionist lead of the Newfoundland legislation, as subsequently did the federal Liberal government with its 1981 National Energy Program. This increased the role of the national oil company, Petro-

Canada (which had been established in 1975 by a Liberal minority government at the behest of the social democratic New Democratic Party), and sought to increase Canadian ownership of the national oil industry, maximise federal revenues, promote Canadian employment and business opportunities and minimise negative social, economic and environmental impacts.

Thus by late 1981 both the Newfoundland and federal governments had adopted variants of the North Sea model, and the industry operated under the regulations of both until the 1985 settlement of the jurisdictional dispute.[8] This settlement gave the provincial government rights of taxation over the resource and provided for joint management by a Canada Newfoundland Offshore Petroleum Board. The legislation provides for both provincial and national preference in employment and business, but there has also been a joint retreat from the interventionist model. The newly elected Progressive Conservative federal government abolished the National Energy Programme in 1985, reducing the prospective role of Petro-Canada and incentives to Canadian ownership and offshore drilling. Similarly, the Newfoundland and Labrador Petroleum Corporation (which was never a functioning entity given the lack of development activity) has been wound up.

As we shall see in the papers that follow, the degree to which the culture of the international oil industry has been modified by the different national cultures, particularly with respect to labour practices, has had significant implications for women. It has affected not only the employment of women offshore, but also the way in which male workers experience offshore work, which in turn affects their wives and children.

WORK AND FAMILY

The workings of the international oil industry and its relationship with different national economies, political systems and cultures is nevertheless only one, albeit crucial, element in understanding the relationship between women as workers and as members of families and offshore oil. We also need draw on our understanding of the way in which the position of women is structured in the developed world.

Recent sociological studies demonstrate widespread acceptance of the importance of the mutual interpenetration of the worlds of work and family, which has come about largely as a result of feminist work in reconceptualising the family and its meaning for women. Leonore Davidoff (1979), for example, showed how difficult it is historically to classify women's activities as domestic or public by taking the example of the work of Edwardian landladies, and Janet Finch (1983) has drawn attention to the way in which the unpaid labour of wives of professional men, such as doctors or clergy, may be 'incorporated' into the work of their husbands. Building on this kind of analysis, Ray Pahl (1984) has sought to draw together the study of production, reproduction and consumption in an effort to reconceptualise the idea of 'work' as opposed to 'employment'. He sets out to do this by focusing on the household rather than the individual and thereby immediately captures the importance of gendered labour in the home and outside it. These and other recent analyses (Brown, 1984; Thompson, 1983; Roberts *et al.*, 1985) accept the homogeneity of peoples' lives, the importance of the 'informal' economy, including the unofficial and unrecognised work women do in the household, and the 'constraints of gender' on men as well as on women.

Such an approach stands in marked contrast to the classical post-war literature, which has represented the family as a private institution that is, or should be, separate from the world of work and public office. This position is common to those who see contemporary family life as having become increasingly privatised to the advantage of both individual and state, and to those who fear that the boundary between private and public is being steadily eroded and that the private world of home and family is being invaded by the public, particularly by agents of the state. The relationship between public and private has thus been conceptualised from the male point of view, with the family portrayed as a refuge to which the male worker gladly returns. Just as the boundary between the two is important to the worker, so its preservation is perceived in most of the mainstream work on the family to be of crucial importance. In so far as women's position has been considered explicitly at all, it is assumed that irrespective of her role in the public sphere, her primary task is to guard and maintain the private refuge of home and family. But the desire to treat family and work dichotomously has meant that the interpenetration between the two spheres, which is crucial to understanding the position of women, has not been made the focus of analysis.

In Parson's (1955) classic formulation, the process of modernisation resulted in the increasing isolation of conjugal kin from the community and wider kin network, such that the family became a specialised agency with the primary task of socialising children to take their place in the wider social system. In Parson's view, the family was to be seen as an harmonious organic unit successfully meeting the needs of industrial society for the nurture of autonomous individuals and as a place where emotions could be given free rein. The increasingly specialised role of the family was paralleled by a clear differentiation in the role of the husband as the externally oriented 'ideas man' and the wife as the internally oriented 'custodian'. Wives might go out to work, but the 'natural' link between childbearing and rearing that was perceived to render both women's tasks would mean that any paid work would be of secondary importance.

In their analysis of marriage published in the early 1960s, Berger and Kellner (1964) went on to stress the way in which the private world became more important in terms of the construction of personal identity as the public world became more complicated. They contended that within the family 'the individual will seek . . . the apparent power to fashion a world, however Lilliputian, that will reflect his own being . . . a world in which, consequently, he is somebody'. Berger and Kellner argued that both men and women constructed their worlds through the institution of marriage in such a way that served to stabilise their realities, but their use of the male pronoun was significant. Some forty years before, Eleanor Rathbone (1936) described a similar process by which men sought identity (and power) in the private sphere, but from a feminist perspective:

> A man likes to feel that he has 'dependants'. He looks in the glass and sees himself as perhaps others see him – physically negligible, mentally ill-equipped, poor, unimportant, unsuccessful. He looks in the mirror he keeps in his mind, and sees his wife clinging to his arm and the children clustered round her skirts; all looking up to him, as that giver of all good gifts, the wage-earner. The picture is very alluring.
>
> (Rathbone, 1936, p.272)

Berger and Kellner saw the continual process by which husbands and wives constructed and reconstructed their worlds through marital conversation as an essentially stabilising and harmonious process and did not consider the unequal aspects of the marital relationship or the possibility of the husband becoming 'a somebody' rather than, or even at the expense of, the wife.

Those historians and sociologists who sought to make the increased privatisation of the family part of the exploration of the changed political consciousness of the manual worker posited a link between the worlds of home and work, but explicitly considered only the male experience. This work marked a new departure in that it took the home life of the male worker to be an important determinant of his attitude towards work and politics, but as Marilyn Porter (1983) has pointed out, no attempt was made to assess the part played by other family members, namely wives, in the construction of working class consciousness.

It has been left primarily to recent feminist analysis to question the whole assumption that the sexual division of labour in the home is natural and unproblematic. The revelations of widespread family violence in the late 1970s served to undermine the assumption that relations between husbands and wives in the late twentieth century were harmonious, and feminist analysis demonstrated the way in which women's position in society was fundamentally structured by their subordination in the family (e.g. Barrett and McIntosh, 1982). In short, men and women experience family differently (Thorne and Yalom, 1982).

From this starting point it has been possible to begin to break down the dichotomy of home and work to understand better the interrelations and contradictions between women's position in the public and private spheres.

First, in respect to work, the majority of adult women in Western countries will take responsibility for a mix of paid and unpaid work in and outside the home. Broadly speaking, the pattern of women's paid work has changed dramatically in western countries, from the 1930s when very few (10 per cent in Britain) women worked after marriage, to the classic bipolar pattern of the post-war years, when married women began to return to the work-force after a period of childbearing and childrearing, and finally to the current prediction that women presently entering the labour market will on average take only a minimal seven year break for childbearing (Hakim, 1979; Martin and Roberts, 1984; Land, 1986). Thus, rather than treating work and the family dichotomously, it is more appropriate to see women's experience of work as part of a continuum. The fact that women must take responsibility for both may be seen to inhibit their opportunities in the field of paid employment. Secondly, the public sphere impinges on women in the family through their husband's work. The construction of private reality, as Burgoyne and Clark (1984) have pointed out, is to a large extent dependent on the income and status of the male earner, which in turn affects the balance of power and patterns of negotiation between husband and wife. In addition, many women will end up with three jobs: paid employment, domestic labour and caring work, and actually contributing to their husband's performance of their jobs. Thirdly, there is the issue as to how far the family is in fact isolated and the extent to which women can rely on the support of other kin, friends and neighbours. And finally, there is the question of how we might achieve change, and particularly the role that state legislation may realistically be expected to play. All of these issues are highlighted in respect to the offshore oil industry, which, with its physical removal from the home and its 'cowboy culture', poses particular problems for women as workers and as wives and mothers.

In Western societies, all work, in and outside the home, is gendered. As Margaret Stacey (1981, p.13) has observed, there have been two quite unrelated theories about the division of labour:

> one that it all began with Adam Smith and the other that it all began with Adam and Eve. The first has to do with production and the social control of workers and the second with reproduction and the social control of women. The problem is that the two accounts, both men's accounts, have never been reconciled.

The boundary between men's and women's work has shifted substantially over time, especially in respect to paid work, but sexual segregation persists. Women now work in a much broader range of paid jobs than they did in the early part of the century, but men's and women's jobs tend to be segregated both horizontally (they work in different kinds of jobs) and vertically (they are found in different grades within the same occupation). There has been considerably less change in the sphere of unpaid work, which remains largely women's work.

Like mainstream work in the sociology of the family, economic theory has tended to treat work in the family and paid work dichotomously. Neoclassicists see women 'choosing' marriage and children and therefore putting less effort into securing well-paid, high status employment. More radical analysis has tended to see men as the passive beneficiaries of a capitalist strategy that divides labour markets into primary and secondary sectors with restricted mobility between the two by a process of deskilling.[9] Men dominate the primary sector where employment is relatively stable and higher paid, and women the secondary sector of often temporary and poorly paid work. However, radical theorists are no more able than their neoclassical colleagues to explain why it is women who occupy the low paid, low status jobs, without resorting to the assumption that their position in the labour market and their responsibility for unpaid domestic work is the natural corollary of their reproductive role. A more adequate theory of women's work requires a broader understanding of the sexual division of work as a part of the construction of masculinity and feminity and of the gender order that pervades all aspects of our economic, social and political life. As Sandra Wallman (1979) has pointed out, work controls the identity as much as the economy of the worker and control of paid work entails not only control over the allocation and disposition of resources, but also implies control over the values ascribed to each of them. The division between male and female work must be located within a structure of male domination which values women's work less. Men, as husbands, employers and policymakers, and indeed women themselves, have always had a strong sense of what work is suitable for women and what work women are capable of. Cynthia Cockburn (1983) has described the process by which men in the British printing trades have identified skilled work as a part of their male identities; women, who are by definition unskilled workers, have been systematically excluded from the trade since the mid-nineteenth century. We might expect the oil industry to provide another case where most of the jobs would be considered unsuitable for and beyond the capacity of women. The vast increase in women's jobs since the Second World War has come in the human service field, in work which has characteristicaly moved between the family, the informal and the formal labour market (such as caring for the elderly). It is not surprising that one of the few areas of offshore work that has been penetrated by women is that of catering.

In the case of married women, the ideal of the family wage has long assumed women's place to be in the home and has therefore prioritised women's unpaid domestic work. The concept of the family wage (which sees men as breadwinners and women and children as dependents) emerged in the nineteenth century and has been fundamental in determining the mix of paid and unpaid work performed by adult married women (Land, 1980). For its acceptance has ensured that women have taken primary responsibility for home and family and that their paid work has been regarded by society and by themselves to be of secondary importance. If married women work outside the home, they are expected to retain responsibility for home and family, which means that many seek part-time work and/or work that is close to home. Given the sexual division of work in the home, it is highly unlikely that many married women would be able to consider work offshore and the rotational shift pattern that goes with it.

Married women whose husbands work offshore must shoulder the whole responsibility for home and family life while the man is away. This makes it more difficult still for this group of women to engage in paid work of any discription. On the other hand it may be argued that they are able to assume a position of greater independence and authority within the private sphere. However, from their study of fishermen and their families, Thompson, Wailey and Lummis (1983) concluded that 'the mere absence of men [does not] necessarily bring women increased independence or respect'.

Women's work is doubly gendered, first being largely confined to 'feminine' tasks, whether paid or unpaid, and second, being subordinated to men's work both at home and in the work place. Thus, Janet Finch (1983) and Hilary Callan and Shirley Ardener (1984) have shown how a wife's labour may be elicited to support that of her husband by incorporating her into the structures around which the husband's work is organised and directly incorporating her labour into the work he does. (The contributors to Callan's book focused primarily on the wives of executives, a group whose experiences are not considered here.) Men who work a rotational shift away from home will make special demands on their wives' unpaid labour, both while they are away and when they are home. Their wives may in turn look to kin and neighbours for support. But while much recent feminist analysis has stressed the importance of female networks, and there is evidence to support the idea that these are effective in certain contexts (Porter, 1984), Bujra (1978) is doubtless more realistic when she concludes that female solidarity rarely overrides conjugal solidarity and rarely results in substantial empowerment.

It is important to see women's experience as being 'of a piece' and the chapters that follow will make clear the extent to which it is so. The question of changing the nature of the female experience is obviously vastly complicated. The chapters on women as offshore workers explore in particular the idea of using affirmative action, which is especially important when it is realised that family law and social policy have traditionally operated to bolster the ideal of a male breadwinner and female dependants. But as consideration of the position of women as the wives of male offshore workers makes additionally clear, ensuring female access to the public sphere of work will have little impact unless first the work culture offshore is changed, and second, sexual divisions are redrawn to ensure greater male participation in the private sphere. Feminists currently find themselves facing renewed pressure from theorists of the New Right who seek to reinforce women's duty to home and family and to use familial rhetoric to attack women's rights. It is precisely the enforced conceptual *separation* of the idea of women as gendered individuals and as mainstays of the family that has enabled such theorists to enunciate positions that are inherently damaging to the freedom of women. Miriam David (1983, 1986) has described their aim to return sexual and social relations to the family as the pursuit of a new 'moral economy'. In accordance with this view, the care of children and of the elderly or disabled are considered to be properly matters for 'the family' (that is women). 'Preparation for parenthood' is to be taught as part of the school curriculum; there is no commitment to expanding women's opportunities in the labour market (indeed in Britain maternity provision has deteriorated and it has been made more difficult for mothers to draw unemployment benefit); while current debates over *in vitro* fertilisation and the provision of birth control information to teenage girls serve first to reinforce the idea of the centrality of motherhood and second as a vehicle for reconstructing the ideology of the 'good mother' as full-time nurturer. However, even in Britain, where New Right ideology holds the greatest sway among the countries under discussion in this book, the clock has not been turned back. For example, the number of married women working there continues to rise. It is proper to recognise, however, that women's choices are at present in danger of being constrained rather than enlarged in the manner that we are advocating here.

Notes

1. The term 'oil industry' is used here, as is common, as a shorthand for the oil and gas (or petroleum) industry.
2. For an overview of the international oil industry, see Blair (1976) and Sampson (1975).

3. This will only occur, of course, where the local culture is more sympathetic to the full and equal involvement of women. This may not be the case in some Middle Eastern states, for instance.
4. This, and subsequent Canadian costs, are converted at an exchange rate of $(Canadian) 2.1 to £1.
5. The North Sea data presented refer to the northern North Sea, including the area between Scotland and Norway, and around the Shetland Islands. They ignore the shallower waters of the southern North Sea, where drilling commenced around 1960, with the first British gas field being discovered in 1965 and entering production in 1967.
6. A number of major oil and gas fields have been discovered in Atlantic Canada since 1979, and prior to the mid-1980s price slump both the industry and governments expected development of the Hibernia oil field (off Newfoundland) and the Venture gas field (off Nova Scotia) to proceed in the second half of the decade.
7. On Canadian federal/provincial relations with respect to the oil industry see Laxer (1983).
8. It is not possible to establish to what degree provincial and federal policies were successful in increasing local and national involvement in Newfoundland's offshore oil industry. However, by mid-1982, 78 per cent of the 1668 jobs in drilling and service operations were filled by Newfoundlanders. Petro-Canada (contracting out to a variety of drilling and supply boat companies) and an Alberta-based joint venture, Husky-Bow Valley (mostly using its own rigs and supply boats) have been the most active operators in recent years.
9. For the various theories of women's position in the labour market, see Amsden (1980), Beechey (1983).

Part I

Women, Work and Offshore Oil

Introduction

THINKING ABOUT WOMEN, WORK AND OIL

We have already discussed the intrinsic and complex interrelationships between the accepted dualities of 'home' and 'work', and have pointed to our growing understanding of the ways in which all aspects of life in patriarchal society are gendered (de Beauvoir, 1946; Gilligan, 1984; Spender, 1982).

The papers in this section, which focus explicitly on the experience of women working, or trying to work, in the oil industry, help to develop certain theoretical issues relating to the sexual division of labour. The oil industry is a traditionally 'male' industry associated with heavy manual work and a tough macho image. In this milieu workers are not only 'obviously' male but exemplary men. Heen describes something of the culture clash when management inbred with such ideas recruit a work-force, like the Norwegians, with very different ideas about how work is organised and why. All this is compounded when the location is 'offshore', at sea, another sphere traditionally deemed to be male in both ethos and practice (see Anderson and Wadel, 1972; Acheson, 1981). It is almost a contradiction in terms to even imagine women in such paradigmatic situations of masculinity. This notion of the work as peculiarly male is highlighted in the neanderthal attitudes of British employers described by Wybrow, but it is also clearly visible in the discussions of the drill floor in Anger *et al*. and Heen. On the other hand, and in sharp contradiction, the oil industry is the cutting edge of new technology. The traditional appearance of work on the drill floor is a front for an immensely sophisticated and diverse technology: one that is changing and developing constantly. This aspect of the oil industry makes it of particular interest in its consequences for the sexual division of labour. In a recent book devoted to the way technological change affects women's work, Game and Pringle (1985) point out that while the allocation of work to men or women is constantly changing, the basis for the attribution is usually held to be the superior biology of the male, and it always has the consequence of reinforcing power relations between men and women. In a situation of constant flux, the only thing that remains fixed is that there is a fundamental distinction between men's work and women's work. The reproduction of these distinctions between male and female spheres, they argue, 'accounts for the so-called naturalness of it all' (Game and Pringle, 1985, p.16). The oil industry provides us with a powerful, technologically innovative example in which we can study the renegotiation of the sexual division of labour and the recomposition of power relations. It is at this point that we enter the realm of the explicitly social, and factors such as legislation, family and marital patterns, trade union priorities, the structure of the labour market and efforts by organised women become directly relevant.

However, study of the oil industry, like others, confirms that beneath the visible levels of social practices lies the ideological underpinning based in gendered consciousness.

As theoretical understanding of such ideas develops, feminists have turned from the close examination of the way in which domestic labour was construed and how it related to capitalist production (see Kaluzynska, 1980; Molyneux, 1979) to look at how women's assumed and actual roles in the family constitute an area of exploitation precisely because of their unacknowledged contribution to the work of either husbands or the state. Much of Delphy's work on the domestic mode of production has explored the theoretical consequences of this view of marriage. Working from within different theoretical constructs and in a different context, writers on the operation of peasant and peripheral economics, such as Deere (1976), have made substantially the same points, exploring the volume and importance of women's economic contribution and the ways in which it is rendered invisible and (usually) appropriated by their male relatives.

Such studies help us to understand the particular adaptations families, and especially women, make to men's offshore work. Solheim's description of the three social realities, 'his offshore life, her single life at home and the joint life of togetherness' is a case in point. These studies simultaneously clarify some of the reasons *preventing* women working offshore *and* those adaptations women make, which enable them to pursue careers in the oil industry. The consequence of such studies is to return us once again to the interrelationship of home and work.

More recent work has, instead, taken up the gendered nature of the work women *do* do. One example of this is the way in which Beechey's (1985) study of part-time work relates to her theoretical suggestions as to why women are always paid at below their value (Beechey, 1983). Similarly, Walby's (1983, 1984) examination of redundancy and women shows how the labour market is constantly restructured and renegotiated on the basis of women's place within it, which is relatively independent of their position in the family.

Work such as this, which tries to separate out what it is that is specific about women's work, rather than how they come to it, and with what constraints, is still relatively undeveloped, and the following papers should provide a contribution from one especially interesting industry.

In terms of the three regions under discussion in this volume, the situation is affected not only by differences in the cultures of Norway, Scotland and Newfoundland, but also in the stage and degree to which the oil industry has developed. The Norwegian example is, in many ways, the most instructive because Norwegian women appear to have been the most successful both in penetrating the industry as workers and in adapting its schedules to their family lives. This process has been helped by conscious efforts in terms of legislation and government encouragement of sympathetic practices within the oil industry. Heen's account, therefore, provides us with the most developed example of the reconstruction of gender and work on offshore installations. As we have discusssed in the General Introduction, it is important to remember that the Newfoundland industry, in particular, is still entirely engaged in exploration, and is never expected to have an oil industry on the scale of Scotland or Norway.

One obvious expression of gendered work is the segregation and segmentation of labour forces by sex. In all three regions under consideration, while women's labour force participation has risen and continues to rise, women are even more concentrated today in relatively few jobs, and those tend to be less skilled, lower paid and with fewer opportunities.

In Canada in 1981, half the women in the labour force were concentrated in 10 leading occupations – led by stenographers and typists (10.1 per cent) and salespersons (6.4 per cent). In Newfoundland, 57 per cent of working women are to be found in clerical work, service industries or sales. In Norway women constitute 45 per cent of those employed in banking, insurance, commercial services and property management, and 65 per cent of those employed in public, social and private services (Norway, Equal Status Council, 1984). Norway has devised innovative programmes aimed at encouraging employers in extremely segregated industries to employ workers of the other sex. For instance in 1979 the Government established a programme under which it paid the first six month's wages of women hired by employers in industries where sexual segregation was particularly pronounced. Since 1985 the reimbursement has dropped to 25 per cent of their wages and benefits. Despite such measures, and continual pressure from the 'Ombud', Norway's labour force remains the most segregated of the three.

The most noticeable aspect of women's work offshore is how traditional most of it is. Yet again, women are concentrated in jobs as cleaners, cooks, nurses and secretaries, continuing a sexual division of labour that assumes mothering and its extensions to be 'women's work'. The 'caring' jobs on the rigs are the first to be handed over to women – catering, nursing and secretarial work. Additionally, the material from both Norway and Newfoundland shows that women offshore are also expected to take on unpaid work as carers. They listen to male workers' domestic concerns, provide psychological support, create a domestic and 'homely' atmosphere, improve the standards of hygiene and cleanliness, and also perform as sex objects, whether in joking relationships (see Chapter 2) or simply as something to watch: 'it's like you're the entertainment', (see Chapter 3). As one offshore installation manager said, 'the first thing that hits you when you get out of the chopper when there are women in the rig is the smell of after-shave' (*Offshore Engineer*, June 1985).

Interestingly, while the women who do work on the rigs and platforms may find some of this additional role burdensome, none of them find it remarkable. They shift imperceptibly from being ungendered 'workers' to being 'women', with all that that entails in their own culture. But no one questions why women are cleaner, more orderly and gentle. Reading the women's accounts of life offshore raises the strong possibility that the intense pressure for women to be acceptable in a hostile work culture actually forces them to overconform to the norm in order to counteract their transgression of it in being offshore at all (for a more general discussion of this point see Franklin, 1985).

The numbers of women who work in non-traditional jobs offshore (as opposed to those working in traditional areas) are too few to permit useful generalisations, but their experience of both the discriminatory hiring practices and the male work culture support the contention that the gendered division of labour is not merely a commonsensical or convenient device, but a fundamental organising principle of our society. Heen's paper makes valuable comparisons between different groups of women workers – secretaries, nurses and catering workers (all traditional women's occupations) and women working in male occupations. It is this last category that experiences the full force of entering a totally male work culture. Heen describes the difficulties involved for those women in creating positions for themselves and women *and* as equals at work. The precarious compromise they achieve is constantly undermined by the contradictory nature of their two roles.

A key aspect of the organisation of work offshore, and one which reinforces its gender specific nature, is both the length of time workers spend away from their homes, and the way that time is organised. Rotational patterns vary. One week on, one week off is common in the Gulf of Mexico and is the norm for some United Kingdom platforms. Two weeks on, three weeks off is common in Norway. Three weeks on, three weeks off is the standard pattern on rigs in Newfoundland and Nova Scotia. On top of this rotation, most workers offshore work twelve-hour shifts. If the shift pattern changes in mid 'hitch' then the problems of adaptation are heightened. These rotation and shift patterns have been seen as one of the principal areas of strain for male workers and their families (a question discussed in depth in the next section) and as a major barrier to women entering work offshore. The absolute separation between work and home is perceived as making it impossible for women to fulfil their dual domestic and work roles (and, to be fair, for men to fulfil their domestic obligations). As Heen demonstrates, women workers in Norway have successfully tried a number of strategies to adjust to and use the rotation to their own advantage (see Chapter 5). In Scotland and Newfoundland there are still too few women involved to make a judgement but the evidence in Newfoundland points the same way. The rotational system also works against the hiring of women because of the twenty-four hour nature of the work on the rigs.

The pattern of working and the highly differentiated and hierarchical organisation of work all make it 'obviously' more difficult for women to combine domestic responsibilities with careers offshore. But many women who work offshore are single or without dependent children. Presumably, therefore, it is no harder for them than for equivalent men. Yet there continues to be resistance to women, any women, working offshore. Why?

Different employers, or the same employers in different places, use different strategies to exclude women, but beneath the querulous wailing about accommodation that some use as an excuse not to hire women lies a more profound disquiet over the confusion of 'home' and 'work'. The intensity of opposition indicates a deeper than rational desire to keep the spheres separate. In this connection it is interesting to note the collapse of the initial hostility of male offshore workers to women on Norwegian platforms, most of whom now *prefer* to have women present (see Chapter 2), and the reason given is nearly always that it *does* break down the separation of 'home' and 'work', i.e. that women 'equal' home and bring it with them in all sorts of tangible ways. If this is the case, then, the introduction of women onto the rigs and platforms is predominantly 'female' occupations simply pushes the home/work boundary out to sea. It in no way challenges or even renegotiates it, and, indeed, makes it even harder for the women who are trying to enter male occupations.

Understanding the roots of gender division in patriarchal society is important in itself. It is also vital to the attempt to alter the specific factors that prevent women from renegotiating the sexual division of labour.

BARRIERS AND CHANNELS: LEGISLATION AND WOMEN'S WORK OFFSHORE

There is evidence, especially from Norway, that cultural or community expectations do change in those communities with more experience of women working. We know that patterns of work and child care can and do change as circumstances change, and resolute attempts to make education and training more encouraging do have an effect. We can be more precise in documenting discrimination at the point of hiring. The best study of this in the oil industry is the work done by Wybrow and Moore in the United Kingdom. This can be supplemented by the more positive experience of breaking down overt discrimination in Norway, where women have been employed in offshore installations since 1977. They first entered traditional jobs as nurses and caterers and later became secretaries and clerks. There has been a quite marked annual increase in the number of women working offshore. By 1984 there were 7700 in the industry, or 13 per cent of the total work-force, and 4 per cent of the offshore work-force. They now make up an substantial proportion of the workers offshore in these traditional 'female' jobs. It has been much harder for women to penetrate to the highly skilled, managerial or traditionally 'male' jobs. However, with the help of legislation and constant pressure from the government (especially the Ombud) it has been possible for 780 women to take up such work (Hagen, 1976).

Experience in Newfoundland has not so far been encouraging. The Environmental Assessment Panel for the Hibernia Project recommended that women's opportunities offshore should be increased both through government policy and project-specific employment requirements (Hibernia Environmental Assessment Panel, 1985). It should be noted, however, that the federal/provincial legislation governing the way in which Hibernia will be exploited only specifically insists on special provision for Newfoundlanders, not for Newfoundland women. In conditions of high unemployment and a culture that discourages women from going offshore, the oil companies have ample scope for avoiding effective affirmative action.

We have emphasised that while the general contours of the sexual division of labour and of women's experience remain fairly constant in the societies we are dealing with, cultural, political and economic differences may profoundly affect both the working of and the renegotiation of women's lives. This is also true of the legal framework. All three countries – Britain, Canada and Norway – have an array of legal inhibitions of overt discrimination on the grounds of sex. Each is embedded in different ideas about the role of law in social reform and about the operation of the law. Each is also embedded in deeply patriarchal understandings about gender relations.

The Norwegian laws appear to offer the best protection for women, coupled with some real efforts to ensure that women have an effective means to pursue their rights. This is rooted in a greater acceptance of women's rights in Norway, and by Norwegian oil companies. Section 1 of the Equal Status Act (ESA), which was adopted in 1978,

> shall promote equal status between the sexes and aims particularly at improving the situation of women. The public authorities shall facilitate equality of status between the sexes in all sectors of society. Women and men shall be given equal opportunities for education, employment and cultural and professional advancement.

A commentary on the Act published in 1985 at the end of the Norwegian Government's two-year Action Plan for Equal Status between the Sexes, went further by saying, 'the law is also intended to influence attitudes to sex roles, committing the authorities to work actively for equal status through instruments which are not encompassed by the Act' (Norway, *Norwegian Equal Status Act with Comment*, 1985). This attempt to influence attitudes and put responsibility for the aims of the Act on to the authorities rather than on to individual women was reinforced by the establishment of an Ombud and an Equal Status Appeals Board. The Ombud's task is to ensure that the provisions of the Act are followed. He/She acts *ex-officio* on her/his own initiative, and also deals with cases brought to her/him for consideration of discriminatory practice. Despite a small staff and limited funding the Ombud deals with more than six hundred cases a year.

In Britain, the long fight for equal pay (the subject of a Royal Commission Report in 1949) culminated in the Equal Pay Act (EPA) of 1970, which was not implemented until 1975. A Sex Discrimination Act (SDA) was also passed in 1975 in the realisation that equal pay legislation alone would lead to the exclusion of women from better paid jobs and thus to an increase in sexual segregation. The Act prohibits discrimination on grounds of sex in appointments, promotions, dismissals and redundancies, access to education, training, and to credit and other services, and includes indirect discrimination where a condition or requirement is imposed on one group so that more of one sex can comply with it. However, unlike its Norwegian conterpart, the British Government has not promoted equal status between the sexes during the 1980s, nor has it made any attempt to extend equal opportunities legislation to the continental shelf. Unlike Norway, neither the EPA nor the SDA cover offshore installations.

Until recently, women seeking redress under the British legislation have found the lack of a unitary piece of equal status legislation (such as exists in Norway) a problem; for example, a female secretary claiming equal pay with a male colleague doing 'similar' work found it impossible either to bring an action under the EPA because the man's work was deemed insufficiently similar (he was not primarily a typist), or under the SDA because it did not cover pay. In some measure this problem has been eased by the amendment to the EPA to include 'equal pay for equal value' (already the essence of Norwegian equal pay legislation) which was made with great reluctance by the Government at the beginning of 1984 in order to bring British practice into line with EEC regulations. Legally, at any rate, it can no longer be assumed that women's work is necessarily inferior, or that women's needs are less.

The results of the British legislation have been relatively disappointing. There was some lessening in both pay differences and the degree of sexual segregation in the work-force between 1975 and 1977, but by 1979 both had returned to 1973 levels. In any case it is arguable that such improvements as there have been in women's pay are attributable either to general economic trends or to greater unionisation among women (Bruegel, 1983). Fewer than one tenth of the applications made under the legislation are upheld; in 1980 only fifteen cases taken under the SDA were successful. Furthermore, in contrast to Norway's Ombud, the British Equal Opportunities Commission (EOC), which works on a pitiful budget (the 1985/86 budget was £3.4 million) fails to use even the powers it has to full effect. If it suspects a breach of law it can conduct formal investigations and can compel organisations to give evidence, but in fact the EOC has completed only a handful of formal investigations, preferring to use 'voluntaristic' methods backed up by research and publicity. In the face of determined employer opposition, trade union apathy and worker resistance, this is simply not enough, as is clearly evidenced by the continuing inequality of women in the United Kingdom labour market. The experiences detailed by Wybrow show how this operates in the oil industry.

The major weakness of both the United Kingdom and the Norwegian approach is their focus on specific legislation directed at particular groups. In Britain, for example, sex discrimination legislation is paralleled by the Race Relations Act. This approach constitutes each group as a unique 'problem' and erects a complex mesh of never quite comparable legislation and case law. It also reinforces the deeply held notion that there is a 'norm' in English law, as everywhere else, that is male, white, Christian and of British ancestry and everyone else is not quite 'normal'. Liberal concerns may force small concessions, but they in no way overthrow the deeply prejudiced basis of English Law, and indeed of British society, that denies the concept of equality based on 'human' rights.

In contrast to Britain, the legal strategy in Canada has not been to enact specific legislation addressing specific discriminatory practices, but to incorporate women in general Charter of Rights provisions and Human Rights codes. The most important federal laws that pertain to discrimination in employment are the Canada Act and the Constitution Act (1982), (the Canadian Charter of Rights and Freedoms) Section 15 (1), which did not come into effect until 1985. This generally prohibits discrimination on a number of grounds, including sex. Section 28 (16) specifically guarantees that all rights and freedoms apply equally to men and women. No cases have yet been brought under Section 15, although a number of organisations are gathering money and momentum to exploit its potential. The Canadian Human Rights Act 1976.77 c 33, addresses discrimination in employment more specifically, insisting that all employment decisions relating to hiring, promotion, training and dismissal should be based on genuine occupational requirements. At the provincial level there are separate (and different) human rights codes and very varied legislation over matters that fall under provincial jurisdiction (which includes labour relations, health and safety regulations, minimum wage levels, specific regulations for domestic workers and regulations concerning provincial public services).

This would seem to provide quite a strong legislative framework, especially when it is taken together with provisions for affirmative action programmes and various protections for maternity benefits, maternity leave and other factors that directly affect women's participation in the labour market. However, the Abella Commission (Canada, 1984) set up to look at equality in employment for four groups – women, native people, disabled persons, and visible minorities – rapidly focused on the fact that discrimination means 'practices or attitudes that have, whether by design or *impact* the effect of limiting opportunities'. So the question of whether the discrimination is intentional is irrelevant to the fact of inequality. But current provisions *are* confined (except in the Ontario Human Rights Code and the Canada Human Rights Act) to cases of *intentional* discrimination, thus immediately limiting their effect. The legislation is also, as in Norway and Britain, damagingly restricted to individual allegations. These are difficult and expensive to bring and have a minimal impact on the structural underpinnings of discrimination. As the Abella Commission puts it,

> Resolving discrimination carried by malevolent intent on a case-by-case basis puts human rights commissions in the position of stamping out bush fires when the urgency is in the incendiary potential of the whole forest.
>
> (Canada, 1984, p.8)

Couching legislation in human rights terms reflects a profound Canadian concern with individual rights and freedoms. By the same token, it makes it blind to the structural causes of inequality and reluctant or unable to address it. Redress at the individual level can never do more than bring a series of 'token' cases to public attention. It provides a formal equality, but it does nothing to attack the real causes of that equality. That is the reason why, despite the strong looking legislation in Canada, so few actual cases have been brought and won under the provisions of the Charter so far.

Affirmative action is a concept that appears to be more sharply directed at structural inequality. In Canada there are legally enshrined provisions (especially at the provincial level) for establishing affirmative action programmes on either a mandatory or a voluntary basis. The mandatory approach has been little used in Canada (e.g. preferential hiring programmes for war veterans after the Second World War) and voluntary programmes have been greeted with distinct ambivalence, especially after their failure to affect the composition of the public sector work-force in the programme attempted after the 1970 Royal Commission on the Status of Women. The Abella Commission is frankly skeptical of such an approach, given the 'intractability of employment discrimination' (Canada, 1984, p.10).

Norway's approach has been more direct. In the oil industry, for example, the Ombud's direct intervention resulted in the Minister of Oil and Energy bringing further pressure on offshore operators to implement the ESA, in both the spirit and the letter of the law, and further inducement is provided by the equal status grants, whereby the state contributes 25 per cent of the wages and costs of workers for the first six months of their employment. Norwegian feminists are cautious in their evaluation of such programmes, but they certainly go well beyond what is available in either Canada or Britain.

Legislation can be strong, but it will be ineffective if it does not apply where it is needed. In the case of the offshore oil industry, the United Kingdom Sex Discrimination Act (1975) is useless because it does not apply to installations on the continental shelf, although there are provisions that would allow it to be extended. The Norwegian ESA, on the other hand, 'explicitly extends to installations and devices on the Norwegian part of the Continental Shelf'. Furthermore, it outlaws reference to the conventional excuses of accommodation problems and lavatory facilities as reasons for not employing women.

Another difficulty with enforcing such legislation is the different statuses and modes of operation of permanent installations, highly mobile drilling rigs and supply boats. Large oil companies have a vested interest in public relations, but the drilling and other subcontracting companies may have much less reason to be concerned about national sensibilities and much greater freedom to avoid legal requirements. It was for this reason that Norwegian unions campaigned to have Statoil take over the catering contracts on its offshore rigs (see Chapter 2).

The difference in the legal practices and the social understandings in which they are rooted help to specify the different circumstances in which British, Newfoundland and Norwegian women enter (or are refused entry into) the offshore labour market. It illustrates the care we should take to relate the experience reported in these papers to the economic, social and cultural communities in which they live. However, it is also clear that despite the differences, women in the offshore oil industry are encountering common experiences. We have discussed the extent to which these may be attributed to something 'special' about the oil industry. While some features, such as the rotational shift system, clearly have particular consequences for women, the papers reveal deeper processes that underlie both the oil industry and the societies in which they operate. These processes are rooted in the patriarchal nature of the societies in which the women of Scotland, Newfoundland, and Norway all live. The practices of the industry may be modified because of technological or economic need or as a result of social pressure, but the labour process will contrive to retain a basic sexual division of labour that depends on a separation of home and work and the continuing domination of men over women in all social relations.

1 Equal Opportunities in the North Sea?

Peter Wybrow

In 1983 the United Kingdom Equal Opportunities Commission (EOC) approached the Sociology Department at Aberdeen University and asked them to carry out a study of the problems encountered by women seeking offshore employment. It had received a number of complaints from women geologists who felt that they had been discriminated against when they had applied for offshore geology jobs. The scale of the problem, if indeed there was one, was unknown. Similarly it was not known whether other groups of women, for example engineers or catering workers, were being discriminated against in the British sector of the North Sea, or whether the situation was different in other offshore areas, such as the Norwegian sector. The report *Women in the North Sea Oil Industry* (Moore and Wybrow, 1984) examined female employment at the two ends of the employment spectrum, catering and geology. Interviews were carried out with the major oil companies and service companies, including the catering contractors. In addition a postal questionnaire was used to survey 161 female geology graduates who had graduated between 1981 and 1983. This chapter describes the results of that study, first comparing the British with the Norwegian experience; secondly, examining women's employment, principally in catering and geology, from an industry perspective; and, thirdly outlining women's own perceptions of discrimination.

WOMEN IN THE NORWEGIAN OFFSHORE OIL INDUSTRY

Like the oil industry itself, the problems of gender inequality are international. In August 1984 the Norwegian oil industry employed a total of 58 732 people onshore and offshore of whom only 7696 were women (13.1 per cent). Female employment began in the Norwegian oil industry in 1979, and the proportion of women employed has steadily increased since then. This is the result of a policy adopted by the unions and government to replace foreign workers with Norwegian, and to a lesser extent, Norwegian equal status legislation (which as we shall see, is much more effective than its British equivalent in respect to the offshore oil industry) and a general increased awareness of women's employment problems. As Figure 1.1 shows the total female employment in the oil industry has increased at virtually the same rate at which the number of foreign workers has decreased.

In August 1984 17 464 people in the oil industry worked offshore in the Norwegian sector of the North Sea. This represents less than 1 per cent of the total Norwegian workforce although the economic importance of this work is far greater, contributing directly almost one fifth of the Norwegian Gross National Product. Women filled 682 positions representing 3.9 per cent of the total offshore

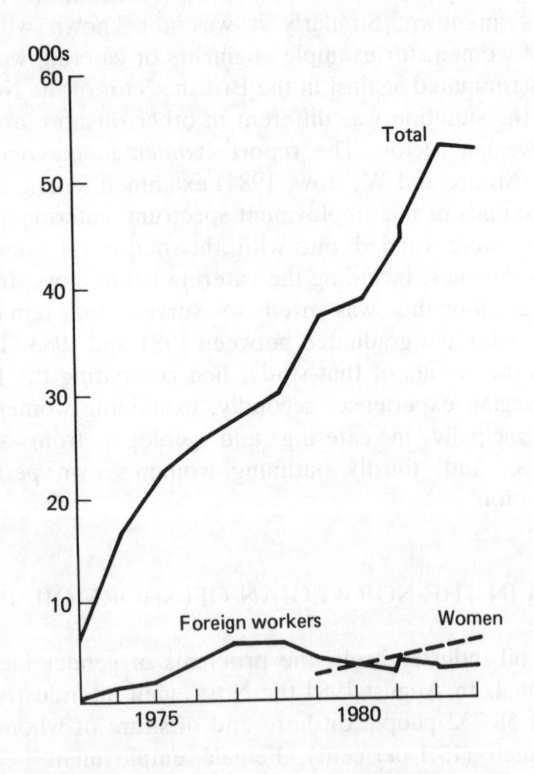

Figure 1.1 Oil employees: North Sea Norwegian sector

workforce. However the proportions of men to women employed varies from field to field and platform to platform. According to research conducted by the Work Research Institutes (WRI), for instance, 80 per cent of Statfjord Field catering workers are women. This represents 10 per cent of the entire field's population.

The growth of employment opportunities for women offshore in the Norwegian sector has been slow despite a history of employment of women in the merchant marine. Women had for many years been employed in the Norwegian ships as radio operators and later as medics. The majority of women employed offshore work in catering although some work as medics, secretaries and radio operators. Only a very small group of women work as engineers, geologists or in other skilled occupations related directly to production.

Figures produced by the Norwegian Work Directorate (*Arbeidsdirektoratet*) for August 1983 provide a detailed breakdown of all the employment in offshore Norway between men and women (Table 1.1). Just over one-third of the 1100 catering staff (6.4 per cent of the total offshore work-force) in the Norwegian sector of the North Sea are women. Wage levels in catering, which in the British sector is regarded as the lowest group in the offshore status hierarchy, are the lowest in the North Sea. Onshore women work predominantly in secretarial and lower administrative levels within the oil companies, engineering firms and the mechanical and shipyard industries (see Table 1.2).

The Mobil research team studying the Statfjord Field (Hellesøy, 1981) have pointed to large differences in males' and females' evaluation of their work environment and organisation. Women are more negative in their evaluations of supervision and support from the work organisation. There are also indications that women receive less on-the-job training than men. Nevertheless, Holter (1984) claims that their evidence, based on study of both the Norwegian and British sectors, conveys

> the impression of a generally positive evaluation among the men of women as offshore workers. Both among catering and other employees, the majority regard the recruitment of women positively. On this score the Norwegian work culture clearly differs for example from the dominant view on the British sector, where there are extremely few females and also a strong male attitude against female employment.

Table 1.1 Male and female employment in the Norwegian oil industry, August 1984

	Onshore			*Offshore*			*Total*		
	Men	*Women*	*Total*	*Men*	*Women*	*Total*	*Men*	*Women*	*Total*
Oil operator	5645	2663	8308	3054	98	3152	8699	2761	11460
Drilling	342	139	481	3374	43	3417	3716	182	3898
Transport	1357	266	1623	4078	21	4099	5435	287	5722
Mechanical engineering	14320	967	15287	3379	26	3405	17699	993	18692
Service companies	893	324	1217	1749	15	1764	2642	389	2981
Engineering companies	4183	1300	5483	182	14	196	4355	1314	.5679
Bases	635	99	734	7	–	7	642	99	741
Catering	16	16	32	661	461	1122	677	477	1154
Terminals	2213	353	2566	27	2	29	2240	355	2595
Construction	2252	246	2498	89	–	89	2341	246	2587
Administration	374	167	541	–	–	–	374	167	541
Research*	1090	171	1261	17	1	18	1107	172	1279
Others	934	303	1237	165	1	166	1099	304	1403
Total	34254	7014	41268	16782	682	17464	51036	7696	58732

* Includes teaching.

Table 1.2 Women in the Norwegian oil industry

Women as % of total work-force	*1984* %	(*1983*) %	(*1982*) %
Industry	13.1	(12.3)	(10.5)
Offshore	3.9	(3.8)	(2.8)
Offshore catering	41.1	(37.5)	(28.6)

CONTRASTS BETWEEN NORWEGIAN AND BRITISH PRACTICE

While the growth of the employment of women in the Norwegian sector of the North Sea is slow, it is nevertheless occurring across the spectrum of companies from the major operators such as Mobil, Elf and Phillips, to the owners of drilling ships and platforms and the smaller contracting companies. Elf had the first ever female offshore installation manager (OIM) in the Frigg field. A number of Norwegian drilling companies have been employing women in various positions for a long time. Two of these firms are of particular interest because they operate drilling vessels in both the British and Norwegian sectors of the North Sea. Both began as shipping companies and diversified into the drilling business as a result of developments in the North Sea.

The first, Company L, owns nine drilling vessels. They employ women mostly as catering helpers and medics and one vessel has a female first engineer. In 1981, the company decided to set up its own catering organisation because they found that contract catering was expensive, of not very good quality and suffered from a very high turnover of staff. Stability of staffing was seen as being important not only in terms of cost and quality, but also in terms of communality, producing a better sense of identification as being part of the crew. The personnel manager for Company L felt that having women working on their vessels produced a 'more natural environment'. It was no great change for the company since, like other Norwegian shipping companies, they have long employed female radio operators and medics, with 50 per cent of the latter female.

While women's stability of employment was not as great as men's, there was some evidence to suggest that this was not due to the nature of the job but to its geographical location. For example, if a vessel moved from the North Sea to the coast of Africa, women would prefer not to move from the North Sea where they could commute home more regularly.

In May 1983, Company L chartered its newest rig to a British oil company to carry out drilling in the British sector, North West of the Shetlands. The rig operated under the British flag, had a British crew and, unlike all other vessels in Company L's fleet, a contracted catering crew as required by the British operator. According to Company L's health and safety manager no women were employed on the rig in British waters.

The second Norwegian shipping company is an old-established Norwegian business which owns seven drilling rigs. Out of a total offshore workforce of 900 they employed twenty-eight women on rigs. These consisted, in January 1984, of two mates, five medics, one radio operator, two cooks and eighteen galley hands. Like the previous Norwegian company, Company K decided in 1981 to employ their own catering staff instead of using contractual staff.

Company K's crew manager (personnel manager) arranged for the EOC research team to meet with a female rig medic, the assistant crew manager and an offshore installation manager.

All expressed positive attitudes to female employees offshore. The rig medic had worked offshore for four and a half years, mostly on fixed platforms, and had found attitudes towards women very negative in her early days on the platforms. However she stated that she now encounters no problems now because she is a woman (and she too felt that women make for a more 'natural atmosphere'). She found work offshore financially rewarding, exciting and challenging. She could earn approximately 50 per cent more working offshore on fixed platforms, slightly less on drilling rigs, than in a comparable job onshore. The medic felt 'very much part of the crew'. Although she had expressed the view that she didn't 'think there were problems because [she] was a women' the medic did 'feel very welcome *because* you are a woman not because you are you'. She had 'no regrets' about working offshore.

The assistant crew manager who was specifically responsible for employing the women, claimed that they increased stability and communality on rigs and platforms and said 'I haven't found one negative thing about employing women yet'. In his experience rigs and platforms were maintained much more efficiently from a cleaning/catering perspective when women were onboard. At the time of the research, Company K was actively trying to recruit more women, especially qualified cooks.

Explaining the company's attitudes he said that women are 'employed not because they are women but because they do similar work onshore. . . . Nurses are employed according to their qualifications . . . we want women onboard'. He went on to say that although there was a positive attitude to women in the Norwegian sector it was still very difficult for them to move into certain sections of the industry. For instance, it is 'more difficult for a woman to become a roustabout' and 'still very hard for a girl to get into a top job position'. There is no doubt that even in the Norwegian sector of the North Sea the oil industry is still very conservative. Everyone spoken to in Norway commented on the fact that having women on rigs and platforms improved living conditions because vessels were kept cleaner; male employees kept the vessel tidier to start with; male employees improved their personal hygiene and appearance; vessels had a more normal atmosphere.

According to the crew manager most of the women who work

offshore are related to someone already working on the rigs. Many are married to men who also work for Company K. While the company cannot promise to place couples on the same rig, they do everything they can to arrange for them to have the same two weeks offshore and three weeks onshore, give or take a day or two. This was confirmed by the crew lists and rosters that were made available.

At the time of the research, Company K had three rigs working in the British sector. There were no women on rig I while there had been one female medic and one female mate on rig II. Rig III was currently under charter to the same British operator which had chartered from Company L. Previously, while in the Norwegian sector, there had been a female roustabout and female helicopter landing officers onboard as well as female catering staff. However there were no women working on Rig III in the British sector. According to the informant in Company K, the British operator stated, when negotiating the contract, that 'employing women in the British sector (of the North Sea) is *illegal*' (author's emphasis). Senior management initially refused to meet the author and later claimed to be unable to discuss this particular case.

British companies often raised problems concerned with accommodation and toilet facilities as reasons for not employing women. But such matters did not seem to be a problem to the Norwegians. Vessels being used in the British sector without women onboard employ them on a regular basis in the Norwegian sector. Company K's most modern drilling rigs have been built with twin cabin accommodation with separate shower and toilet facilities for each cabin. In the Norwegian sector such twin cabins are shared by men and women on separate twelve-hour shifts.

THE IMPACT OF THE NORWEGIAN EQUAL STATUS COUNCIL (*LIKESTILLINGSRADET*)

The provisions of the 1978 Norwegian Equal Status Act, which aimed to both ensure substantive equality of treatment between the sexes, and to influence attitudes towards sex roles, were explicitly extended to cover the Norwegian continental shelf. This stands in contrast with the 1975 British Sex Discrimination Act, which does not apply to the North Sea. The Norwegian Act established the Equal Status Ombud and the Equal Status Appeals Board (ESAB).

The Equal Status Council (ESC), originally set up in 1972 to promote equal status in family life, schools, business and communities was also reorganised under the Act.

There have been few complaints to the ESAB since it was set up. By 1984 they had investigated nineteen cases of which four concerned refusals by companies to employ women because of accommodation problems. Not all cases examined were the result of women making complaints; in a few cases companies sought a ruling as to whether it was illegal to refuse to employ women on certain specific grounds. For instance a diving company wanted to reject a woman diver for a deep diving job because it involved six people living in a small pressure vessel for three weeks at a time. Within this confined space the six people had to eat, sleep, dress and undress. There was a small separate toilet and a small room between the toilet and the main part of the chamber. The *Likestillingsradet* ruled that it would be illegal to refuse women employment on the stated grounds.

According to the Ombud, Eva Kolstad, relatively few women are seeking employment offshore. As in Britain and Atlantic Canada, offshore jobs are not generally advertised and it is thought that it is mainly women on the West Coast of Norway who have friends or relatives who work offshore that seek such employment. The major operators express an interest in hiring women but, according to the Ombud, they do not actively recruit them. The Minister of Oil and Energy has sent a letter to all the operators pressing them to observe the law and employ more women, but there seems to be a lack of will in implementing the Act. The ESC has no enforcement procedures itself and can only pronounce as to whether the law has been broken or not. In the words of Eva Kolstad: 'the will to give power to the ESAB is not the same as the will to create a beautiful structure'.

In the period 1979–81 the ESAB, with a staff of seven, dealt with 2500 cases and it is constantly under pressure. Eva Kolstad has said that with insufficient staff some of the most time consuming cases have not been handled adequately, while some of the easier have not been dealt with quickly enough. In terms of the enforcement of Norwegian law offshore the ESC finds international companies 'not so willing to accept Norwegian Law, especially American companies'. Multinational companies were also seen as being 'very hard on their employees'. The Ombud believes that the 'oil industry is *much worse* than the onshore industry' (original emphasis) and thinks this is because the ESAB came onto the scene a little late, when an 'American culture' had already pervaded the Norwegian sector of the North Sea and made the task of enforcing equal status more problematic. Indeed, its success has been uneven.

One of the major grounds for not employing women offshore in the British sector – inadequate lavatory and showering facilities – is *legally inadmissible* in Norway. It is incumbant on companies to provide such facilities for both sexes. Some of the companies operating in the Norwegian sector have made major progress in providing accommodation. On the Ekofisk field, for instance, Phillips Petroleum have installed two-berth rooms with separate shower and toilet facilities. At the Board level Phillips have also created an equal opportunities council. The Ombud however felt that the usefulness of this development still had to be seen: 'its fine on paper but in reality we still have to see concrete results'.

The Equal Status Act also makes it illegal to differentiate between the sexes in terms of educational opportunities, but there is evidence that it does occur. The Ombud received complaints from two young women in a Bergen engineering school which provided a three-year course on how to operate and maintain equipment used in offshore drilling. The course consisted of two years theoretical work and one year practical. The women complained that young men were accepted for practical training offshore but that the companies refused to take the women. Their failure to obtain practical experience meant that they would not be accepted at the drilling school, the purpose for which they were training. The Ombud wrote to the engineering school several times over a two-year period but the school failed to reply to any communications. In the end the women also gave up corresponding with the Ombud.

Finally, the rights of pregnant women offshore was a major concern of the Ombud. Pregnant women are required to come ashore at twenty-eight weeks. In Norwegian law they should be given a job with the same pay as when working offshore, but in reality such a job is seldom found for them and they have to go on 'sick pay' at a considerably reduced income.

There is no doubt, despite the application of the Equal Status Act to the Norwegian continental shelf, that discrimination occurs in offshore work. However it has been difficult to prove. In 1980/81 the Ombud received a number of complaints from women wanting jobs as roustabouts. Eva Kolstad reported that her team 'couldn't reach the facts' in these cases and there was no definite conclusion. However, as the process of Norwegianisation has progressed, women have found it relatively easy to move into the areas of work seen as 'women's work' onshore such as nursing and catering/cleaning services. Companies ranging from service companies to major oil companies seem prepared to continue the myth that women are biologically better adapted to provide for the 'domestic' needs of men. They are less keen to promote the interests of women when it comes to employment in areas more directly concerned with oil and gas production.

The ESC has also found difficulty in getting the public to take the question of equal status for women offshore seriously. A female anthropologist researching the effects of offshore work on oil families commented on the 'remarkable silence from the feminist movement' in Norway over the issue of equal status offshore. Thus even in a country where equal status legislation is extended offshore women are poorly represented in the labour force. However it must also be reiterated that there is little evidence to suggest that there are large numbers of women seeking work offshore. Representatives of service companies, operators and the Ombud all agreed that what was occurring offshore was a slow change in culture from one which had been predominantly a male orientated 'American hire and fire mentality' toward a more stable, egalitarian Norwegian culture in which women would take an increasing role.

WOMEN IN THE BRITISH OFFSHORE OIL INDUSTRY

There are between twenty-two and twenty-three thousand employees working in the United Kingdom sector of the North Sea but the number who are women is not known. Official employment statistics are not categorised by sex and this is not surprising given the very small numbers of women involved, while neither the industry itself nor the Department of Energy could provide figures on the numbers of women working offshore. However, a survey of offshore workers passing through the Aberdeen heliport showed 0.1 per cent of the inbound and outbound passengers to be female, although the numbers were so small (fourteen out of 14 000 in one case) that one female catering crew could significantly alter the percentage. However, even if such an alteration trebled the number of women the proportion would still be below 0.5 per cent compared with the Norwegian figure of 3.9 per cent.

At one heliport all the female passengers were asked about their occupation and the duration of their work offshore. Three women (a geologist, a petroleum engineer and a structural engineer) had worked offshore for some time but were only travelling out for three to four days. On their first trip offshore were a secretary (going for two days) and an administrative assistant and computer programmer on day visits. The best estimate based on these figures and on knowledge of the industry was that there were about twenty-five women working offshore on a normal shift basis in the British sector.

DEFINING WOMEN'S WORK IN THE BRITISH SECTOR

Catering

At the time of writing about 22 000 people are employed offshore in the British sector of whom it is estimated that around 2000 are catering workers. In January 1984 it was possible to find only one catering company employing women offshore. They had been approached in mid-1983 by one of the operators to provide, on an experimental basis, eight women to work on a platform. By the autumn of that year the number had increased to sixteen, and the company had plans to further increase the number of women employed. A representative of the catering company stated that they would like to recruit female cooks but that it was difficult to attract female staff of the right calibre. By mid-1985 the number of women employed by this company had fallen back to eight and it was stated that it was not envisaged that this would increase in the foreseeable future.

This company claimed to be the only catering firm employing women offshore in the British sector and this was confirmed by other caterers, although several commented that they were considering employing women in the near future. The company claimed to have approximately seven hundred staff working offshore such that the sixteen women employed represented approximately 2.3 per cent of their offshore work-force and approximately 0.8 per cent of the total British offshore catering work-force. This compares very unfavourably with the Norwegian figure of 41.1 per cent.

The women had reserved accommodation on the platform and worked the same shift patterns as men, such that at any one time there were eight female catering crew offshore. The operator and catering contractor were said to be pleased with the experiment and the representative of the caterers thought that the employment of female catering crews offshore would increase. So far, they had found no negative aspects to employing women. Most of the women were not recruited from the onshore catering business, as might be expected, but were former sales representatives, nurses and teachers. The two weeks on, two weeks off rotation was said to be a major attraction for the women.

It was not possible to talk to any of the sixteen women because the operating company did not want the experiment publicised. Interestingly this was the one operator that refused to grant Moore and Wybrow an interview concerning female employment offshore for the EOC study.

Geology

The EOC study began in a period of declining opportunities for all new graduates in geology, irrespective of sex. Potential employers and university and polytechnic geology departments reported the decline, which was underlined by the withdrawal of at least one major servicing company from the yearly hiring 'milkround'. Even though by the end of the study employers were beginning to report a shortage of trained geologists, the earlier trend had affected the perception of the labour market and job opportunities for present students. From a survey of post-1980 female graduates it was found that about 80 per cent would have liked jobs as geologists but that less than 30 per cent obtained them. It is important to note in passing that interviews showed, first, that many women who had wanted to be employed as geologists were in fact very happy with the jobs they had and were finding their basic scientific training useful in their work, and, second, that some of the women who did obtain geological jobs found them much less interesting than undergraduate work. They thought 'being a geologist' would mean doing research or working in the field when in fact they ended up doing routine work in an office.

It should also be noted that it was not possible to compare the success of men and women in finding work in the oil industry. It was not in the mandate of the EOC study to collect data on male geologists. The purpose was to collect data on the extent to which women graduating in 1980–83 had been seeking work in the oil industry, and on their success in finding employment in the North Sea. Those women who failed to gain employment in the oil industry may have made more or fewer applications and had more or fewer interviews. The question that interested us was whether their lack of success was due to their sex, rather than, say, job shortages. The evidence provided by the EOC study and reported here comes from the author's impressions of interviews, the experience of women already in the oil industry and statements made to them by the prospective employers of geology graduates.

The questionnaire was sent to 195 women who graduated from geology departments between 1980 and 1983; 161 replies were obtained. Only 18 per cent of these respondents were married, while 82 per cent had a second class degree or better. One hundred and twenty-eight (80 per cent) had wanted to continue working in an area related to geology after graduating and 100 of these reported making 1400 applications for oil-related jobs. They received 221 interviews resulting in twenty-nine jobs (i.e. jobs for 23 per cent of those wanting work in geology). Of the twenty-nine who finally obtained oil-related employment, twenty-five reported making 450 plus applications and receiving eighty-three interviews.

Among those wanting work in geology thirty-three (26 per cent) did not apply for particular jobs because they believed women were excluded. Nine women reported that companies declining to interview them had explicitly stated that they did not employ women. Of the women seeking geological employment, thirty-five (27 per cent) reported that they were asked at one or more interviews about marriage and/or personal relationships.

Of the twenty-nine women who succeeded in obtaining oil-related jobs, fourteen had attempted to get work offshore, four had been explicitly denied work offshore because they were female, seven believed they may have been refused a job because they were women. Twelve of the twenty-nine did not apply for specific jobs because they thought women would not be hired. Only four women reported working (or having worked) offshore on a regular basis, one of them in the Norwegian sector. Four others worked offshore occasionally.

It was found that 16 per cent of the respondents had been discouraged from seeking employment as geologists by university teachers and 19 per cent by university careers services, friends and school teachers. In all, 30 per cent of the female respondents had received discouraging comments from one or more sources.

A letter from a schoolgirl during the research showed that discouragement could start very early:

> I am sixteen years old, in the Lower Sixth . . . I am studying Geography, French and Latin at 'A' level, and have recently commenced the Geology 'O' level course. I find Geology a most interesting subject and am keen to pursue it at a higher level. However, I have been most perturbed when making enquiries about job prospects for Geologists. It seems to me that the majority of companies have no interest whatsoever in female geologists. I realise that the Sex Discrimination Act does not apply to underground employment, but I still feel that the whole situation is grossly unfair.

There is also considerable evidence as to the discouraging influence of university and polytechnic staff. Two of the geology department heads who refused to help in our work did so most emphatically: one did not even wish to discuss it, while the other said on the telephone, 'I thought everyone knew that they simply only employ men offshore and will not take women'. He said that he could not see the point of the study and added that he would not have answered our letter. It is interesting to speculate upon the careers guidance this head gives to his female students (or, indeed, if he has any female students).

Final year female geology students, about to begin the 'milkround' in search of employment, were also interviewed. They all commented on the masculine culture of geology departments: the sense in which women have to become honorary men in order to survive as students; the heavy work in the field and the ritual beer drinking with the lads; and the fact that the men did not expect to have to pay any attention to women's interests or problems. However it was noted that honorary male status was less important where the women formed a minority large enough to stick together as a group. Many female students complained that they were not taken seriously:

> In the department you're treated like a potted plant . . . you ask a question, they don't quite say it, but you can see that they mean don't you worry your pretty little head about that. But if one of the blokes asks the same question they get a lot of attention and finish up discussing it in the pub.

One group felt very strongly about staff attitudes, saying that one lecturer (who never actually addressed a woman in his class) stated that women have no place in geology. If a woman volunteered an answer to a question he would reply, 'What a clever girl'. He also offered the opinion that, 'You girls would do better if you didn't spend so much time screwing around.' When being interviewed for this department another student was told that it was a man's profession and was asked if she realised that she would have to become a school teacher. The students in four departments specifically mentioned discouraging and demotivating comments and attitudes on the part of academic staff. One important aspect of the experience of female students is the lack of contact with women professional geologists. Women are encountered as research assistants and geological museum curators, but only occasionally as members of academic staff.

The examples quoted above are of very obvious and rather extreme cases of prejudiced male attitudes. Many other members of staff cared very much for the careers of their female students and tried to give sound advice, although these men often adopted 'realistic' attitudes which stressed the problems women would encounter and tried to protect them from disappointment. The effect of this, without a parallel commitment to helping overcome prejudice among employers and to encourage young women to press for employment, may have been the 'cooling out' of potential professional geologists. Being a caring Head of Department may entail helping a female student to battle against prejudice, but more often seems to mean a paternal and protective attitude which does not adequately serve the interests of educated young women in the 1980s.

In four universities the students picked out the Careers Service as being especially discouraging or even insulting, suggesting careers in nursing or social work to women who wished to become professional geologists.

> The Careers Officer discouraged women from applying for jobs with oil companies since we could be taking up valuable interview time for the men.

> The Careers Officer told me at the beginning of my interview to forget about being a geologist. Then . . . he asked what career I was interested in. When I told him I didn't know, seeing as I was supposed to forget about geology, he asked me what I was doing

using up his time if I didn't know what I wanted to do.

A Careers Adviser at university said of mudlogging that 'You will enjoy it if you like living rough, prostitution and bars'. This was clearly an attempt to put the girls on the geology course off going into logging.

These experiences contrasted with those reported in the two universities with geologists on the careers staff, where the female students remarked warmly upon the value of an adviser who knew the field.

The purpose of dwelling upon these issues is simply this: before discussing oil company practice account has to be taken of the fact that not many women apply for work in the industry. For example one operating company was trying to remedy the lack of women employees. It had recently begun a programme of in-house training of technical staff and tradesmen and had been surprised and disappointed in the response from women, but had nevertheless been able to place one – out of six female applicants – into the scheme. The same company had also wanted to recruit staff for an onshore gas plant and had been disappointed because none of the over 2000 applicants was female. Women do not offer themselves for employment, either because they are interested in other careers, or because they have come to believe that there is no point in seeking work in certain occupations. The point of discrimination may therefore not be in the boardroom, at local management level, or at an interview. A number of the oil companies commented on the lack of applications from women, adopting a liberal stance and alluding to the EOC's Women in Science and Engineering campaign. Women's work expectations are constructed by a complex process in which family, friends, school, universities and polytechnics all play a part.

COMPANY POLICY

While the management of all the oil companies asserted that it was their policy not to discriminate, all of them referred to the difficulties of sending women offshore. The most common problems mentioned were the lack of adequate accommodation and the undesirability of sending women offshore alone. Others cited the practice of contractors, saying that they themselves did not discriminate but that catering companies, for example, tended to use only men – largely because of the problems of rough living and limited accommodation. It will be seen below that the contractors offered a different and at times contrary interpretation.

The availability of accommodation, however, does seem to be an issue when very small numbers of women work offshore. This is especially the case in the development phase of an oilfield, when the accommodation may be temporary, in four-berth units, and often overcrowded with additional temporary berths. There will be contractors and subcontractors on board and a high turnover of personnel, making flexibility in the use of accommodation essential to management. During a discussion with one personnel manager in a firm that was currently developing a field, the telephone rang with an enquiry about forty additional cabins which were to be provided on the platform *that night*. The company was currently using two-berth cabins in sets of two, with shared facilities for the four occupants. This gave 240 berths, but temporary modifications had been made in order to squeeze 410 persons onto the platform. If women were to be accommodated separately and optimum use made of berths they would have needed to be employed in multiples of four.

In some of the earlier developments in the North Sea where jack-up rigs had four berth cabins some contractors would have had to employ all female crews, or to coordinate their accommodation needs with other contractors for women to have separate cabins. However the women interviewed for the EOC study said that they were accustomed to sharing accommodation with men during field trips as students or at remote well sites onshore. With one exception they said they would not mind sharing with men offshore. A mudlogger in the Norwegian sector already did so, and said that neither she nor her male colleague cared who slept in the other bunk – they both had a job to do. However this idea alarmed some British managers interviewed for the study; some feared a *Sun* newspaper headline about 'love-nests in the North Sea', but more were worried about public opinion and the attitude of wives onshore.

Because of the design of new rigs and platforms it is becoming decreasingly the case that the lack of suitable accommodation can be advanced as a reason for not employing women offshore. New platforms are built to specifications that allow use by both sexes, with more accommodation being provided in two-berth cabins with showers. Furthermore, as fields move into production the demands on space are reduced, and permanent 'floatels' may be provided in populous fields.

One British company had, according to a Norwegian engineering contractor, asked to have a rig modified so that it could not be used by men and women. This is the same company that later chartered a rig from another Norwegian company and insisted on all the female crew being replaced by men, giving the reason that women were not allowed to be employed in the British sector of the North Sea.

A different line of reasoning was used when another company explained that it would not be appropriate to employ women offshore in a firm that had mid-western United States origins. Employing women would not be in accordance with the company ethos (though plainly such a policy on the part of the company at home would contravene United States law). In the mid-1970s it was common for employment agencies in Aberdeen to say that many United States companies, especially drilling companies from Southern States, stipulated 'No blacks, no women'. The number of black British people working in the North Sea is not known but is thought to be very small.

The attitudes and policies of the subcontracting service companies are very similar. None of the geological/geophysical service companies approached in the EOC study employed women offshore. In responding they made no attempt at prevarication, and did not assert women did not want to work offshore. The following were typical answers to the question 'How many female geologists do you employ offshore?':

> We don't have any women working offshore.
> We don't have any women geologists offshore at all.
> We don't employ female geologists.
> The answer is very short, we don't employ women offshore.
> We employ no females offshore at all.
> We don't have any women offshore; we never have.

These companies are the major employers of geologists and geophysicists in the North Sea. All had interviewed one or more of the female geologist respondents in the EOC study. Many have claimed to be interested in employing female geologists in the 'milkround' interviews at universities. Indeed a number of these geologists commented in the questionnaires, for instance, that in the 1982 interviews at universities one of the companies made much of the fact that they were hiring four women out of an intake of about one hundred graduate geologists. When this company was asked how many women it employed offshore the answer was:

> A. We don't have any women working offshore at the present time. We did offer four places to women back in 1982 but due to the recession and cutbacks we had to postpone their recruitment. We have written to them recently and 2 or 3 of them now have jobs. We plan to take on some more when we start recruiting again. It would cause us some difficulties because we operate around the world and there are problems employing women in some countries. They would have to be prepared to work in the North Sea. We did get permission from a number of companies, back in 1982, to place women in the Danish and Norwegian sectors.
>
> Q. But not in the British sector?
>
> A. I don't know. I wasn't here then.

One female company representative was more forthcoming:

> We don't employ female geologists . . . it's a taboo subject in this company. We are extremely biased in this company against female geologists. It's something I don't agree with. I'm afraid you would have to talk to one of our directors in . . . about that one. The usual answer is that there isn't enough bed-space on the platforms to employ women.

One of the main reasons given by the geological and geophysical companies for not employing women is lack of physical strength.

> The nature of the work is such that it entails lifting heavy equipment and things of that nature which women simply cannot do. We do, of course, employ women onshore in geophysical analysis, data processing, that sort of thing.

The question of physical strength is also raised with respect to diving. Women are employed in certain limited areas of the commercial diving world. There are female diving instructors at various diving schools and training establishments. There are considerable numbers of women divers in research establishments associated with universities and other research bodies but there are no women divers employed offshore in the British sector. A senior representative of one of the world's largest diving companies explained this in the following way:

> the North Sea is a very male dominated area . . . The problems of introducing one or two women divers would be . . . well it wouldn't be diving that would be the question mark it would be their sex. Not so much what they are doing, it would be very much easier to get over the professional barrier. The sex barrier is a major concern . . . There are certain types of diving, for example inspection work, which are not particularly demanding. I don't think you could use the physical argument there. A much more fundamental problem is that it is a male dominated industry which simply does not cater for women *at all* as a general statement. (Respondent's emphasis)

But even if the problem of physique was overcome there was, in the opinion of employers, still the problem of accommodation, 'if we did have a Tamara Press and we wanted to send her offshore then we'd come up against the problem of accommodation'.

Some of the contracting companies stated some of the oil operators which employ their services were against having women offshore: 'People pay lip service to this sort of thing but the majors are very resistant to change and are very conservative about having women offshore.'

One contractor claimed to be actively trying to get a female engineer offshore:

> We have one female engineer at the moment working onshore. We have approached some of the operating companies to see if she would be acceptable to them for employment offshore. So far they have just ignored us, as if we hadn't asked. This is just my opinion but it seems to be related to the 'youth' of the oil company. The older, longer established companies are very old fashioned and conservative. The younger companies will accept women offshore and there is no problem. It'll come and it should come; there is no difference between a female geologist and a male geologist.

Despite the fact that there was corporate discrimination, not all of the company personnel interviewed concurred with the policy. After saying that his company didn't employ women offshore one of the respondents went on to say

> I have worked on rigs and platforms in the Norwegian sector with female catering staff and medics and there were no problems at all.

> I worked on a rig west of Ireland some time ago and we had a female geologist onboard. The old-hand drillers didn't want to know. She shared a twin-berth cabin with a male geologist, it was frowned on by the older men. The younger men just accepted it. The woman had had to undergo the survival training, etc., along with the rest of the men. In fact she was more highly rated on the survival course than any of the men.

Women are discouraged from applying for offshore work at a number of levels. In their advertising for instance companies describe the people they are looking for as 'men, single, under 30' or they want to hear from 'bright young men' or 'young, single, fit men'. In a letter sent out to University Careers Officers in late 1983, Company W claimed to be looking for sixty graduates for 1984 as follows:

A) *Seismic Field Exploration*	(men)	30
B) *Well Survey Exploration*	(men)	5
C) *Seismic Data Analysis*	(men & women)	25

For world-wide operations.

In an accompanying leaflet the nature of the tasks and opportunities of these occupations are explained:

> *Seismic and Well Survey Parties* (A and B)
> Promotion to positions of high responsibility is rapid for those who, beyond their own specialisation, make themselves acquainted also with all aspects of the work of the field party, and who develop qualities of leadership both of the small European crew of technicians and of the large force of local labour employed in some tropical areas.
> Service is world-wide. On most assignments four weeks of home leave is taken after ten weeks of continuous operations although in certain circumstances in an area such as Australia, home leave may be taken at longer intervals up to a year.
> Camps are frequently in desert, jungle and swamp, and employees on marine surveys live on board ocean going ships.
>
> *Seismic Processing and Analysis* (C)
> Candidates selected for the processing and analysis of seismic data acquired by our field crews will be employed initially at the Company's offices in [the south of England].

> From these new Graduates, who are classified Assistant Seismologist, whilst training, we select staff for transfer to the overseas Processing Units.

Thus it would appear that women will only be considered for less challenging jobs at the head office or processing units overseas, and they are excluded from the field work, on- or offshore, which is necessary for promotion. The women who were not deterred by the written word and had managed to get interviews found that these deterred them or led them to believe that they were not being taken seriously.

WOMEN'S PERCEPTIONS OF DISCRIMINATION

While company representatives listed numerous reasons why they would be unwilling to employ women offshore, there was also often a certain ambivalence about the position of some: either they agreed that in theory there was no reason not to employ women, or that in the future things would undoubtedly change. The experience of the women interviewed for the EOC study was unequivocal. Those reaching the interview stage experienced a variety of tactics on the part of employers aimed at discouraging them from pursuing offshore work. Respondents to the postal questionnaire recorded that:

> I replied to an advert of [Company Y] asking for well loggers. Some time later I received a phone call asking if I would attend an interview, not for a job as a well logger but for a job working in their core analysis laboratory in London . . . At the interview I was told that the company only had one female logger and that she was unattractive, middle aged and incredibly butch. I was told that the working conditions were highly unsuitable and also that the presence of women on the rigs presented a security problem in terms of personal safety. In short he tried to convince me that I didn't want to be a logger anyway. I was offered a job in the laboratory which I subsequently accepted.
>
> When I was interviewed by [Company Y] on the milkround they specifically stated that they did not allow women to do offshore work, although the interviewer said that women were eligible for jobs at their laboratories. I was so disgusted at this that I decided that perhaps geology was *not* the career I wanted after all.

Several women respondents complained that men with inferior degrees were hired in preference to them;

> On being told I wouldn't be physically strong enough [for offshore work] I challenged this and it was admitted that I was as strong as a male of my build of which there were two interviewed the same day – both offered jobs both with 2.ii degrees as compared to my 2.i.

Some women spoke of humiliating tactics on the part of interviewers

> [Company T] told the students that they would not discriminate but informed the head of department that they would not take any women. [Company R] asked 'Could you change a lamp if it broke?' and similar trivial questions that they would not ask a bloke. Most of my interviews were at this level . . . also questions about boy friends and marriage. I was very struck by the [male] clubby atmosphere of the offices in which I was interviewed . . . I was also asked 'would you burn your bra' if you worked in an office full of men?'

Similar trivial questions were also asked in a more bullying style, as this respondent to the questionnaire revealed:

> During interviews with mud logging companies and oil companies, female colleagues were asked questions specifically because of their sex, i.e. do you know how to change a car tyre/electrical plug? Also, they were shouted at by the interviewer to determine how they would react under pressure. Neither of these lines of interview were taken when the same interviewer assessed male interviewees.

Another respondent reported an interviewer using the issue of dress, rather than everyday technical competence or physical strength, to the same effect:

> I was interviewed for a job as a geophysicist. The interview commenced with the Managing Director of the company asking me why I was smiling (I was trying to look pleasant and not nervous), he said, 'I've had girls in here burst into tears because of what I asked them.' He also made a rather critical comment about me having my best interview suit on.

All women applicants, whether single or married were subjected to questions about their personal lives:

> An interview with [Company D] made much of the fact that I would be working in the Netherlands – did this make any difference with regard to relationships? They were told I was unattached – the point was *pressed* – did I think I would *ever* want to get married? I answered 'quite likely' – they went on to question me about my (rather elderly) parents. I explained that my brother lived 5 miles away from them and could cope with any problems (he shows no sign of wishing to move away). The reason given to my Professor for turning me down: doubts about my mobility.

One woman interviewed has succeeded in obtaining a post that would give her the offshore experience crucial to furthering her career. She was well-respected by her colleagues and was head of her section. However, she reported that her local manager was hostile to the idea of women working offshore and insisted upon her taking a chaperone with her, a young typist who had no work to do offshore. (It has been said that a number of young women acting as chaperones make large demands of the women they accompany because of boredom and loneliness.) The local manager was gatekeeper to the North Sea, so although the women geologists felt unhappy about this situation, her need for experience offshore effectively prohibited complaint.

Married women rarely got past interview panels:

> Sheila is married to a company geologist. She was told by one service company that according to company regulations they were *not allowed* to employ women offshore. At all her job interviews she was asked if her husband approved of her working and one interviewer expressed surprise that her husband had *allowed* her to take a degree and then to work. When she was asked what her husband did for a living the results were 'disastrous' at Company G (because of the security risk).

> Fay is married to a geological consultant. She had worked for a geophysical research company but had been given only residual jobs. Her Head of Department said 'We don't want women to improve themselves here' and later when she applied for a newly created post for which she was qualified her head froze the appointment. Company C's Exploration Manager had said he would not employ a woman because other wives would be jealous if a female employee went offshore. She found the same attitude at Company AA. Only one firm, Company BB, regarded gender

as a non-issue, they wanted to be assured that women would take their turn offshore. She is now doing a routine geophysical job but would prefer something more interesting. Sometimes there is a potential clash of interest in data coming to herself and her husband at home, or papers left lying around. Without ever discussing it they found themselves automatically avoiding one anothers papers, not discussing work and going out of the room during phone calls.

A number of academic staff with experience of the oil industry confirmed the attitudes encountered by the women. One ex-Company H geologist said that United States companies in the drilling business were especially chauvinistic – no blacks and no women. One rig manager had said to him, 'If women come on this rig, I'll quit.' Another had asserted that, 'If a woman sets foot on this rig I'll run her arse off of here.'

It has been shown above that many women regard laboratory work as second best to work 'in the field' and yet they are often offered such work when seeking employment offshore. The few women working offshore who replied to the questionnaire clearly confessed their state of isolation and marginalisation, which become all the more oppressive in the context of the 'total institution' of the rig or platform. The woman whose manager insisted she be chaperoned when offshore was told by a tool-pusher when she arrived offshore: 'I've never taken orders from a woman in my life and I don't intend to start now.' He queried every instruction she gave by telephoning his headquarters ashore. They told him to do what the geologist said. This particular incident illustrates a special source of conflict for geologists: the ethos of drilling is to keep going, deeper and faster. Geologists tend to tell them to slow down or stop. Downtime costs money. Thus geologist and driller come into conflict. The drilling crew are said to embody the most *macho* values and to celebrate physical strength and toughness. If the geologist is thought to be a 'slip of a girl' the conflict is amplified.

As one respondent commented:

> I have the feeling that older management in particular, as well as some colleagues don't take women seriously and may not give your point of view as much weight as a man's. We still have to prove we are better at a job than a man, before we are treated with the same attitude and respect. They can't seem to accept that we want a career as much, if not more than they do.

CONCLUSIONS

There can be little doubt that there is conclusive evidence of widespread, almost universal, discrimination against women in employment in the British sector of the North Sea offshore oil industry. While companies accept in principle that increased female employment is inevitable there is very little positive action to bring it about. Moreover the oil industry is, by tradition, a male dominated industry and relatively few women apply for jobs in it.

However, since the publication of the EOC report, a number of major oil companies have been seriously reconsidering their policies in response to equal opportunities. For example women are now employed on a regular basis on BP's Magnus platform. It seems that the tide may at last be set to turn. The increase in the number of women so far seems to have caused the company no problems at all. Indeed it may be suggested that the more women who work offshore the less problems there will be, as is demonstrated by experience in the Norwegian sector.

So far there is little evidence of change on the part of the offshore service companies in the British sector. Most discriminate in offshore work as a matter of policy and say privately that they have little or no intention of changing their practice. These companies none the less continue to interview women even though they have no intention of employing them, partly because publicly they profess an interest in recruiting women for offshore work, and partly because they use the women who apply for offshore jobs as a source of highly skilled labour for onshore laboratory jobs. These jobs, however, offer virtually no prospects of advancement in the industry.

The Norwegian experience shows that women can be employed offshore in a wide range of jobs without adverse effects on the women themselves, the work environment, or the work-force. The women employed in the Norwegian sector are heavily concentrated in traditional 'women's occupations', cleaning, cooking and caring for men. Nevertheless, they are also more extensively employed than British women in managerial and technical occupations. Similarly, Anger, Cake and Fuchs show, in this volume, that the Newfoundland offshore industry (with a total work-force roughly one-tenth that of the British sector) employs as many women as are employed as in the British North Sea.

The failure of women to apply for offshore employment in the British sector is only in part a function of their exclusion. It is also the outcome of a life-long process which discourages girls from following a scientific or technical education and which lowers their educational aspirations and their expectations of employment in the best paid and most technically challenging work. Parents, schools, universities and colleges bear a responsibility for this state of affairs. Whether the women seeking employment are highly educated female geologists or relatively low-paid catering workers, they are denied this opportunity because they are women – *and for no other reason*.

2 Making Out in a Man's World: Norwegian Women Workers Offshore

Hanne Heen

INTRODUCTION

The labour market in Norway, like everywhere else, is highly segregated according to sex. This is the case if we look both at the type of work women do, and where in the work organisation they are found. Women are confined to a limited number of occupations, where they make up a large majority of the employees. For instance, 95 per cent of nursing work is done by women, 78 per cent of service work (including 87 per cent of hotel, restaurant and domestic work) and 77 per cent of clerical work; but they do only 20 per cent of administrative executive work (Norway, Central Bureau of Statistics, 1983). Women are also mostly found in low status positions. In 1981, 7 per cent of women and 28 per cent of men had top jobs.

When offshore oil activity started in Norway in 1966, offshore work was an exclusively male activity. The first women started to work offshore late in 1978 or in the beginning of 1979; these women were probably the first women offshore workers in the world. There are still very few women working offshore – 4–5 per cent of the offshore labour force in the Norwegian sector of the North Sea is female, but the numbers on different platforms varies considerably from zero up to 10 per cent. Women offshore are mainly found in catering, but they also work as nurses, secretaries and radio operators, and a few are employed in traditionally male occupations like roustabouts, operators or as engineers. There was also one female platform manager in the Frigg field, who has now advanced to be manager of the offshore hook-up of a new production platform.

The sexual division of the labour market is generally explained by women's and men's different roles in the family, and/or by the forces operating within the labour market itself (Hartmann, 1981, Furst, 1985). Women's weaker position in the labour market is seen as reinforcing their role in the family and vice versa. Expectations about what women can and ought to do is also important for understanding women's position in the labour market.

The debates about employing women offshore have also concentrated on women's position in the family and on the suitability of women for offshore work. Commonly held beliefs about the nature of the platform society and the physical and psychological strength needed for working there have been raised as arguments against employing women offshore. To admit women into an all male environment also triggered off fears about women's destructive influence, particularly in respect to the anticipated effects on sexual behaviour. The actual experience of employing women has shown such fears to be groundless; instead of creating conflict and dissatisfaction, the presence of women is now perceived to have created a more lively and 'natural' atmosphere.

In addition, women's role in the family has been seen as a major obstacle to their offshore work. It was argued that even if it were possible for single women to work offshore, it would be impossible to combine offshore work with having a family and raising children. From a survey done in Statfjord 1980, we know that of 75 women then working there, 43 per cent were single without children, 16 per cent were single with children, 25 per cent had a partner and no children and 16 per cent had both a partner and children (Hellesøy, 1981). Thus, even though almost half of the women were without family responsibilities and only 32 per cent had children, these statistics show that it is *possible* for women to combine offshore work with having a family.

In what follows, I will describe different groups of women working on the fixed, integrated platforms on the Norwegian Continental Shelf, and analyse their situation in relation to the characteristics of the platform society.[1] Both the description and analysis are based mainly on field studies on the Ekofisk and Statfjord field, and home interviews with male and female workers.[2]

OFFSHORE ORGANISATION

Differences Between Drilling Rigs and Integrated Platforms

The centres of offshore activity in the North Sea are floating rigs which are usually made only for drilling, and fixed installations which usually undertake both drilling and production. On the drilling rigs, there is a single, dominating area of action: the drill floor. While drilling, all activity on the rig is subordinated to this activity, and the drill crew is the dominant social group. The social relations and work culture in this group are therefore of prime importance for the relations and work culture on the whole rig. Even if the degree to which its members subscribe to the traditional male values of hardiness and strength is often exaggerated, it is nevertheless an exclusively male group. In practice the work of drilling is closed to women, and very few women apply for jobs on the drill floor. This is due partly to the physical strain of the work, and partly to the perceived male exclusivity of the group (see also Chapter 3). This means that on the drilling rigs the place where the 'real action' takes place is totally closed to women.

On the integrated platforms, production is the main economic activity, and production work is executed by the employees of the operating company, which is responsible for running the platform. However, only about half or fewer of the people working on the installation are employed by the operating company. The operators' employees work mostly on monitoring production, and in general maintenance work. Drillwork, catering, construction and some maintenance work are done by contractor firms, which are contracted to the operating company for different periods of time. This division results in a difference in social status on board, whereby those employed by contractors find themselves in a subordinate position compared with those employed directly by the operating company.

The production platforms on the Norwegian continental shelf are also predominantly inhabited by men. As with the drill rigs, platform work has traditionally been associated with male virtues and has been associated with hard and long working days in tough surroundings. The few women offshore workers are confined mainly to traditional women's work, but the possibilities for women to expand their areas of work, ought to be larger on the fixed installations with their greater variety of tasks, than on the drilling rigs. Indeed, this is acknowledged by the Norwegian State's oil company, Statoil, which has a policy of recruiting more women for offshore work.

Production Platforms as Social Systems

Drilling for oil on the Norwegian shelf started in 1966. The first oil deposit considered profitable was found in 1969 and production started in 1971. When the oil industry came to Norway, United States oil companies were invited to develop the first fields. They brought with them an American work culture and designed the platform organisation according to their own traditions. This tradition may be roughly characterised by authoritarian leadership; a concept of unions or other forms of workers' organisation as subversive; extensive hiring and firing; and a career system based upon personal achievement, but also upon acceptance and subordination to the system. Most of the leading personnel were then America, and few Norwegians saw any possibility of advancing to any position beyond that of work group leader.

This work culture differed greatly from Norwegian work traditions, where industrial relations are usually regulated through agreements between unions and employers. In the first years there were many conflicts offshore due largely to the clash between two different traditions of work. Over time, the situation has 'normalised' from the Norwegian point of view, due to the imposition of official regulations, increasing unionisation, and the Norwegianisation of leading personnel.

The operating companies also learned to accept that their usual way of running platforms was not necessarily best suited to the Norwegian situation. Thus life on Norwegian platforms today is more firmly rooted in Norwegian culture, and its values and forms of behaviour are easily recognisable as Norwegian. But these traditions are nevertheless grafted onto a foreign work organisation, which appears to Norwegians to be old-fashioned in terms of its adherence to adversarial work relations. The distinctive pattern of development in the Norwegian industry further means that our discussion of Norwegian platform life is not necessarily valid for other parts of the world.

The gradual Norwegianisation of the platforms has also resulted in the progressive employment of women offshore, with women being hired to replace Spanish catering workers. Even though their numbers are few, their impact on the social environment has been disproportionate. To have women on board is seen as significant in itself, independent of the nature of their work. When there are women present, the platform is experienced as having both a more 'natural' and a more lively atmosphere. There also seems to be stronger limitations as to how far extreme forms of male attitudes and behaviour are allowed to develop. The men show more concern about their personal appearance and, perhaps most important, the difference between social life on the platform and at home is significantly reduced. Women are considered to be particularly valuable as catering workers. They are said to add a 'female touch' to the work which makes the living quarters more home-like. The men also tend to keep the living areas tidier when catering work is done by women, and it seems that men more easily accept reprimands about untidiness when it comes from a woman.

Before women went offshore, male workers were very suspicious of the idea of employing women, and the first women were sometimes met by open hostility. These days are now gone, and today most offshore workers say they *prefer* to have women on board. However, many men are still reluctant to actually work alongside women in their own work group. In these circumstances women are perceived to be more threatening, and to have a potentially disruptive influence on the companionship between men.

The reaction encountered by the first women going offshore, was similar to that experienced by the first women working on merchant ships. The acceptance of women on board ship was also seen as a significant step, and was held to have changed the whole atmosphere on board. Korbøl (1970) argues that problems perceived as emanating from the presence of women on board, were in fact related more to the workings of ships as total institutions. When women arrived on board, any difficulties were perceived to derive from their presence, and by projecting the problems onto the women, the strains between the men could be lessened. However, there are important differences between platforms and ships, which account for the more rapid acceptance of women on offshore oil platforms than on ships. One of the main differences is the length of the stay. The long time on board ship exacerbates social tensions generally. This together with the ready acknowledgement given to men's (if not women's) sexual frustration, puts additional stress on the women workers, especially when their numbers are small to start with. On larger oil platforms the women in catering are sufficiently numerous to form groups which can become an independent social force, an impossibility for the few women on a ship.

However, despite the significant changes that have taken place, all offshore platforms are still heavily dominated by men. Thus the primary reference point for an analysis of women's position offshore, must be the nature of the interaction between women and a male work culture. The roles that exist for women offshore are limited, with regard to both work and interpersonal relations. The women meet an offshore culture made and dominated by men, where being female is often seen as their primary characteristic, even where this is modified by the nature of their work.

Work Organisation

The platforms on the Norwegian shelf vary greatly in size and in the number of work processes that go on, from the small gasboosters with a population of around twelve people up to Statfjord and the Ekofisk Complex where 500 people may be present at any one time.

Different types of work go on simultaneously. The flow of oil and gas and the processing must be controlled and managed, maintenance and catering work is done regularly. The organisation of work is done largely through different work groups, each of which performs a specific type of work. Work groups are coordinated through a traditional, hierarchical organisation which has many levels of command. The work is usually quite specialised, although not necessarily highly skilled, which means that there may be little variation in the content of many jobs. Due to difficulties in coordinating the work of different groups, the work day often includes a lot of waiting time. Jobs are also frequently interrupted, partly because emergencies arise, partly because of organisational difficulties. The result is that for many groups, little professional pride is invested in the work; instead it becomes more important merely to get through the day.

Every platform employs people from many different firms. Because much of the work is put out on contract, the people working on a single platform have different employers and their allegiance to the platform varies considerably. This adds to the organisational complexity of the platform.[3] Big and relatively stable contractor groups with a defined task to do, like the drilling or the catering crews, constitute fairly independent work groups with few connections to the operator's work organisation. Other contractor groups, like construction or maintenance groups, are more dependent upon the workers of the operating company, and the difficulties experienced by the operators' personnel in planning and organizing, then spread to the contractor groups that depend upon them. The contractors thus tend to suffer even more interruptions in their work, and to experience longer waits to perform their specific tasks. As many contractors work on the same platform for only a few trips, their workers seldom get the time to develop relationships with the stable population on the platform.

The platform should be seen as a twenty-four-hour society or a total institution as well as a work place. While everyone on the platform is there to work, it is also a place to *live* for a period of time. This not only means that necessary biological needs like rest, sleep and eating must be attended to, but also that the social environment on the platform will encompass a variety of different social settings and social relations found onshore. At the same time most social relations on the platform are overlaid by the conditions and relations originating from the organisation of work, and land based social relations are hard to replicate offshore. All platforms therefore offer an environment with little social stimulation compared to life onshore.

The work group as well as being fundamental to the working of the platform is also the most important social group. The exact meaning of the term 'work group', or 'our shift' as it often is called, may vary, but usually it refers to the people who work together most of the time. As the working day is twelve hours, most of the day is spent with members of this group, and it thus becomes the basis for most leisure activities. The social cohesion of the group is in large part dependent upon the work organisation and upon the way in which the work schedule is set up.

The work schedule on the Norwegian shelf is built upon a ratio of 2:3, with two-fifths of the time offshore, three-fifths of the time onshore,[4] although the way in which rotation is organised varies. In some cases the whole group changes at the same time, in other cases only half the work group change at any one time. The stability of the work group is therefore by no means certain because of the repeated change in personnel. This also increases communication problems between shifts, and when combined with the risks and complexity of the work, makes highly structured job descriptions and work organisation necessary.

Offshore Life

The great variations between tasks, groups, administrative hierarchies and firms means that it is impossible to give a general description of what offshore life is like. An attempt will nevertheless be made to describe some typical characteristics.

One of the most striking aspects of offshore life is its monotony and dullness, springing from the limited possibilities for social activity on the platforms, and from the restrictions which in practice constrain the development of deeper social relations. Boredom is a common refrain of almost all personnel on the platforms, and the less the variation and challenge in the work itself and in the social relations surrounding it, the stronger the perception of monotony and dullness. Tedium is also reinforced by methods of work organisation which render much of the work routine; the jobs at the lower end of the work hierarchy involve little decision making or planning. This organisation of work, combined with the amount of time spent waiting and the routine nature of much of the work, make it important for the workers to develop ways of filling the time. Yet at the same time there are many constraints upon the development of more fulfilling social relations. The offshore milieu fosters a sort of personal reserve. As very little actually happens, small incidents may create much talk and many rumours. Reserve and a conscious lack of commitment to the platform community become ways of avoiding gossip, but such behaviour also makes it difficult to create a wider variety of social relations and impedes the development of closer personal relationships. In consequence, a number of people experience loneliness on board, a feeling that is only intensified by the number of people who are always around.

To have women on board automatically broadens the spectrum of possible social relations and consequently reduces the monotony. The presence of women not only has consequences for relations at the individual level, but may also change the whole atmosphere on the platform and relations between the men themselves. The women have considerable opportunity for social interaction with the men and with each other, but because they are so few and because everybody watches their behaviour, they are especially vulnerable to gossip. Their behaviour on the platform is also interpreted differently from onshore. Very small incidents, like a conversation running a little too long, can be the source of extensive rumours. Consequently, most women restrict their behaviour accordingly.

The impact the women have on the social life on the platform depends on how many there are and on the job they do on the platform. The formation of their own social relations is in fact dependent on the interaction between their sex roles and their work roles.

THE WOMEN ON BOARD

I will describe four groups of women offshore doing different forms of work with the intention of investigating how characteristics of occupation and gender interact to create specific work roles in the social environment of the platform.

The Women in the Administration[5]

This group only exists on the bigger platforms. It consists of a fairly stable small group of women. Most of them are recruited from the oil company's onshore office from women who wanted to get an offshore job. Their work offshore consists of taking care of bookings for helicopter flights, providing a reception service, allocating beds, secretarial work for the administration, and solving a variety of unexpected problems for people all over the oilfield. Work on the reception desk is a reasonably autonomous job, while secretarial work is more closely tied in with the rest of the administration's work; the women rotate between the two types of work.

This work is typical of women's work onshore. The main difference between working offshore and onshore is therefore not found in the nature of the work itself, but in the fact that most of it has to be done immediately; decisions can seldom be postponed. It is possible to see this work as 'processing work', with one of its main tasks being the monitoring of the flow of personnel. Flexibility and personal kindliness are important qualities, as is the strength of character to stick to a decision when necessary.

On Ekofisk, the women in the administration work in the Complex, which is the administrative centre for the whole Ekofisk field. They are classified as part of the administration, which means they work with the top of the field hierarchy. At the same time their own position in the hierarchy is low; their salary is low compared with the rest of the operator's personnel, and their possibilities of advancement offshore are almost zero. As with female white collar workers onshore, there is no career ladder from their work into other parts of the organisation. However, the dead-end nature of their jobs is exceptional within the context of the offshore organisation as a whole.

The work of secretaries onshore is embedded in traditional sex roles, whereby female subordination is mixed with formal subordination within work hierarchies. Secretaries' work is supportive in nature and in the case of personal secretaries, their status is closely linked to that of the man they are working for. They also often have to perform domestic duties in the office and are perceived as playing the role of the 'office wife' (argued by some to be a natural source of job satisfaction for women).[6] Offshore the situation is somewhat different. The work of the secretary is supervised and defined by men, but secretaries do *not* act as supporters for specific men. However, both onshore and offshore the work they do is defined as being outside the general career system, and their experience is almost impossible to convert into other areas of work.

While the position of the women in the platform administration is thereby subordinated to that of the men, as a result both of the nature of their work and of gender relations, there is also room for considerable ambiguity. Their work is perceived as complementary to that of male administrators, and this serves to modify their subordination as women. To be a woman means that they are defined as having qualities which men do not have, and this may render their *formal* subordination less total than would be the case for a man. Women often perform tasks and accept work relations which would be degrading to a man, but which are not necessarily perceived as such for women. Women can be considered 'separate but equal', both by themselves and by others, even if in reality the overall consequences of this 'separate but equal' status means effective subordination.

Relations between men and women are therefore seldom clearcut relations of domination and subordination. By using gender to legitimate domination, the man at the same time puts his own masculinity to test, something which may prove hazardous in the context of closed platform society. It may serve to limit his exercise of authority over women workers even though relations between the sexes are undoubtedly defined on male rather than female terms.

Joking relationships,[7] which are common between sexes on the platform, can be seen as one manifestation of the ambiguity in the relations between male and female workers. The content of the joke usually plays on the sex of the participants. The interaction between those taking part in the exchange may resemble flirting, but there is usually no possibility of it becoming serious, which is what characterises 'real' flirting. The joke often contains references to the man or the women's relationship or attitude to persons of the opposite sex, and to the possibility of a relationship between the participants. To develop a joking relationship, the partners have to accept each other as sexually desirable persons. By accepting the terms of the joke, the woman appears to accept the social premises set by society for relations between men and women, based on female subordination, but the joking relationships also underlines the complementarity between the sexes. It may be difficult not to enter a joking relationship, especially if this type of behaviour is common in a particular work environment (as it is offshore) and is held within accepted limitations. Nevertheless, the woman *can* refuse to participate. As the joking is done in public, this may be taken by the man to constitute a public refusal of himself as a man. One consequence of this is that a joking relationship will only develop between people who are on quite friendly terms with each other. The women will not accept jokes by men they do not like, because they then become unpleasant comments rather than jokes. By rejecting jokes women also make an implicit statement about the sexual desirability of the man.

Joking relationshps are one important way of handling ambiguous social relations and of easing social tensions. For a total institution also to be a work-place is in itself ambiguous: usually one type of behaviour is expected from work-mates and another from people who are a part of one's private life. On the platforms both sets of rules and relationships must be negotiated by the same set of people. Joking relationships and social withdrawal should be seen as different ways of handling the doublesided nature of social relations on the platforms. Such relationships create a special form of closeness, a 'pretended' intimacy that does not run the risk of growing into something else. Jokes manage to express closeness and distance at the same time, the very characteristics that dominate platform society. Joking relationships are common all over the platforms, but more so in heterosexual discourse, where both the need to maintain distance, and the tendency to closeness, are greater.

The women in the administration also find it easier to develop other forms of informal social relations with the wider cross-section of the platform population than do the men. This is partly because they meet almost everybody in their job, but also because they are seen as marginal members of the administration, not least because they are women. The women's position in the administation is thus profoundly ambivalent. They are seen by others as a part of it, and they have access to much information which is concealed from other groups. On the one hand they may have some decision-making powers over personnel, for example, whom to shuttle. On the other hand, their position in the administration is clearly a subordinate one. To be female mediates the ambiguity between the low status of being a secretary and the high formal status of the administration. To be female is also to have another base for interaction with the rest of the platform population. Everybody knows them and they are welcomed by most other groups. However, even if their status in the administration is low, their status is high compared to many men, and clearly higher than the status of the women in catering, which also means that they are accorded a certain amount of respect.

The Nurses

There is a nurse on every platform and on the bigger ones there may be two or three. They deal with cases of illness and injury and have the authority to send people home for medical reasons. Their main medical work is limited to first aid, the diagnosis and preliminary treatment of illness, and the treatment of the many minor discomforts which are common offshore. The nurses' work also includes controlling the standard of the water supply, hygienic standards, and the work of the catering crew. However, the work of the nurses usually extends beyond purely medical questions. They often take on the role as confidante, counsellor, comforter and social worker on the platform. To understand the position of the female offshore nurse her professional role must be considered in relation to both her place in the work organisation and in the offshore social environment. Her role as a professional makes her position different from most other women on board. She is the medical authority, and can decide matters of medical relevance. She can send personnel onshore if she deems it necessary; even if she then usually confers with a doctor onshore, nobody else on the platform can demand to know why. She is outside the regular work organisation, and on most platforms she has no colleagues. Most, if not all her social relations on the platform will be coloured by her special position, for the role of nurse means something to all workers on the platform.

The offshore environment is potentially stressful. It is characterised by the need to keep distance in socially crowded surroundings and many workers experience considerable boredom as well as anxiety about gossip and rumour that spread easily. Because social relations are extensions of work relations, they are very limited. The social environment offshore leaves little room for emotional life and close emotional relationships are quite exceptional. Such relationships provide support in times of stress, but offshore, where the need for emotional support may be considerable (both as a result of the situation on the platform and with regard to the family life onshore), such relationships are almost non-existent. There is therefore a considerable degree of psychological stress inherent in offshore work, and very limited social mechanisms to enable the individual worker to cope.

It falls mainly to the nurses to diffuse the situation. Many people turn to them to talk about personal matters and even when people do not articulate personal distress, nurses must still take the totality of the offshore environment into consideration in the course of their work. The nurse must therefore maintain a balancing act between a strict professional role on the one hand and a more informal stance on the other, whereby she engages in relations characterised by just that confidence and care which the platform society lacks.

The precise nature of this balance depends on the age, sex and personality of the individual nurse. The tendency of the nurse to extend the work role beyond the strictly professional aspect, is larger in female nurses than in males, and also larger in older than in younger nurses. We find that female nurses easily develop a work role characterised by caring and comforting, both for physical discomforts and in cases of psychological stress. Men more readily expect care and emotional support from women than from other men, and women are more often prepared to give it.[8] However, the social circumstances of the platform make the relationship between the nurse and the rest of the personnel different from traditional relations between the sexes. The fact that the relationship is defined as one where the nurse gives (professional) help, and the other person receives it, serves to limit the relationship.

It is important that the nurse can be trusted. The other person knows that she is bound by professional secrecy, and he can be sure that she will not betray his confidence. The nurse's role as 'confidante' easily grows out of this. As a professional, the nurse is also in a situation where she can hardly refuse to talk to anybody. In so far as the other person manages to define the situation as one in which he seeks her help, she is obliged to do her best to help whoever approaches her. A person seeking her aid, does not, therefore, run the risk of being rejected; the nurse is obliged to care. On the other hand, her acceptance of the worker as a patient and her willingness to listen, does not oblige her to any further personal involvement.

The nurses' work extends to the whole platform population. While the majority of workers perceive all time offshore as work-time, it is more continuously work-time for the nurses than for most other personnel. This means that the way in which the nurse manages to develop her work role is of critical importance for her well-being on the platform. Being a nurse gives her a good base for establishing relations with the rest of the platform population, based on a form of respect that cannot be commanded by other female workers. If, however, the nurse does not manage to achieve a working balance between the professional and the more informal contact expected of her, she will easily become isolated and may also be perceived as arrogant.

The Catering Group

The largest group of women working offshore is that of the catering workers. They are mostly employed as service workers, in jobs that demand no formal education. The service workers perform cabin service, work in the kitchen, dining room and laundries, and are responsible for cleaning and tidying the recreation areas. The work is divided into tasks which are the responsibility of one person. To distribute the work between the workers is the responsibility of the steward, but usually the workers do the same work on each trip.

Formally there is no division of tasks between male and female service workers, but in practice women are preferred for cabin work. The quality of women's work here is judged to be higher than that of the men and is perceived to be a function of inherently female qualities. This may be interpreted as one example of refusal to evaluate traditional female competence as an occupational skill, worthy of extra reward. Moreover, because the ability to keep up a high standard of work is perceived as a sex-specific rather than a socially constructed skill, any man who does this sort of work well runs the risk of calling his masculinity into question. Thus men may need to keep a distance from such work in ways that women do not have to, which also means that low quality work by male cabin staff easily becomes a self-fulfilling prophecy.

The catering group is different from other work groups on the platform in a number of ways. The work itself is to take care of the personal needs of the platform population. If we compare the platform to a home, catering work is analogous to the work of the housewife. This is also how it is often perceived by the catering women themselves. The catering group is usually perceived to 'belong' to the platform, and yet be significantly different from the other work groups on the installation. Their work is necessary for the running of the platform, but is not directly connected with the oil activity as such, and it has no direct economic impact. The status of the catering group is generally low in the platform society, due to a number of factors: it is done by contractors, whose workers are usually of lower status than the operators' personnel; it has no direct connection with the oil production; and it tends to belong to the category of reproductive labour which is generally poorly rewarded and poorly regarded.[9]

However, the low status of catering work probably imposes more of a strain on the men in the group than on the women, as we have also seen in the case of women in the administration. To do service work is not degrading for a woman, but it may be for a man. This gives rise to differences in terms of the relations between the men and women inside the catering group and in their relations with the rest of the platform population, beyond those produced by simple gender difference. To be a female caterer is to have a firm base for interaction with people outside the work group, while to be a male caterer is often to have a more ambiguous position than that experienced by other male workers.[10]

The social composition of the catering group differs from the other groups on the platform. The workers are usually younger, a larger proportion of them are single and the sex ratio is more equal. This is also a group where the status of the women is not systematically subordinate to that of the men, which has consequences for the relations inside the work group.

The development of shared standards are important in catering work in respect to the work itself and to patterns of behaviour. Failure to fulfil work standards is usually blamed on the catering group as a whole, and behaviour that crosses accepted norms will, in the case of women easily be blamed on all women. Consequently we find the exercise of strict self-imposed social control inside the catering group, particularly in relation to the maintenance of standards of work and codes of behaviour between the sexes. Sexual relations may develop between male and female workers and be accepted, but only if stable. Anything that is perceived as 'loose' behaviour is much frowned upon and condemned.

The standards of work are often set by women in the group, especially by women who have been working offshore for some years. Inside this core group we often find quite close relationships. These sometimes extend to the women's pattern of social contacts onshore, something that is more common among female than male workers. In male–female relations within the catering group, greatest weight is attached to companionship and equality. Relations between the men and women very often appear a-sexual, despite all the joking.

The contact the catering personnel has with the rest of the platform population differs according to the type of work they do. The service workers who work in the recreation areas, dining room, coffee bars, or who bring food around, easily get to know lots of people, while those working the cabin service see few people in the course of their shift. However, most catering workers usually take their coffee breaks at the table next to the kitchen door in the messroom. The messroom is a public, or at least a semi-work public area outside mealtimes, and on large platforms someone usually sits at the catering table most of the day. The catering group is often more lively and seemingly more open than other groups on the platform, largely because of its different social composition, and it acts as a focal point for other workers, especially those waiting to do a particular job. The nurse is also often seen with the catering group. She does not need to spend all the day in the office, and the catering group is near at hand and inside the living quarters when the nurse needs company.

Generally, the number of women in the catering group is an important factor in determining the success of their adaption of offshore life. When their numbers are few, women seldom constitute a 'core group' in the catering team, and their control over both work standards and social realities is less. Their opportunities for developing companionship with the men outside the group is also lower, and in these circumstances we find that the women tend to keep much to themselves.

Women in Traditional Male Occupations

There are few women in traditional male jobs offshore, but they are nevertheless found in many different occupations if the Norwegian sector as a whole is considered. However, they are scattered around on different platforms, and are usually employed as the only woman in a particular work group.

The women in traditionally male occupations are in a totally different situations from the other women on board. By going into these occupations they not only advertise themselves as competitors in the game for advancement, but they challenge the intimate connection between these jobs and masculinity, not by attempting to feminise the work, but by implicitly stating that there is no direct connection between occupation and gender. This may be threatening to male workers and may be one of the reasons for the male workers' reluctance to accept women workers as equals. Male hostility often expresses itself in terms of misgivings as to whether a woman can do the job well enough, or complaints that she is being given an unfair advantage, for instance in getting lighter work. If the woman works hard and gains promotion, she may be told that her advancement has come more easily because she is female and that she has not actually deserved it.

Women tend to play down the negative reactions they get from men, saying that such reactions are the exception rather than the rule. However, what they do say is that the men have to be 'handled right'. They feel that it is their responsibility to make themselves respected and to achieve this by doing a job as good as anybody else, and by refusing to play the role of sex object. This means that they must reach a balance between agreeing to be treated 'as a woman' on the one hand, and as 'one of the guys' on the other. If the balance tips in favour of the former, then they are in danger of not being taken seriously. This might also mean being placed in the role of 'mascot'. The ideal seems rather to act and be treated as 'one of the guys' in the sense of being accepted as a 'full member' of the group, but not in such a way as to deny femininity. The problem is dealt with by sidestepping the problem of female subordination or the 'secondness' often implicit in being treated 'as a woman'. The decision about how she is treated does not, in the final event, belong to the woman worker, but she can by her method of handling relations and situations, exert some control over the way in which she is perceived. Her task is difficult because she must create for herself a position as a woman *and* as an equal worker, for which there are few role models. She has to create a qualitatively new role, in which to be female is not to be subordinate but is still to be different. The strategy most women follow is to emphasise their role as workers in their relationships, in other words to emphasise the characteristics that make them similar to the men. They do their best to define their gender as a personal characteristic; they are female and want to be accepted as such, but they refuse to accept gender as the most important factor defining their relations. An important consequence of this is that they seldom develop relations with other women on the platform merely on the basis that they are the same sex.

FAMILY RELATIONS OF OFFSHORE WOMEN

Women with family responsibilities who work offshore must make a major break with commonly held perceptions about women's role in the family, because one of the main family responsibilities for women is to *be* home. For women to be able to leave the family regularly for about two weeks at the time usually requires a restructuring of family tasks and responsibilities. The implications of this vary according to the age of the children, and whether the woman is married.

Few women with small children work offshore: the children are perceived as needing their mother and the mother seldom wants to leave them. As the children get older, the mother feels freer to leave them. However, this does not necessarily mean that her absence is unproblematic. It is not only children who expect care from women in the family, but husbands also claim their time in respect to both household labour and companionship. Many women perceive regular periods of absence as being incompatible with their role as wives.

When the mother is single or divorced, the situation may be very different. If the woman can find satisfactory child care while she is away, either by her divorced husband, or by her own family members, she will often reckon work offshore to be preferable to work onshore. Offshore work makes it possible to have full-time work with reasonable pay and a concentrated period of free time at home with the family. It is possible to be a full-time housewife when at home, and the woman will have considerably more time to spend with her children than if she worked full-time onshore.

If the woman is married and has children, her husband will usually take on the responsibility for the children while she is away. As most men work outside home, this means that smaller children must have another caretaker during the day. This is also the case when women work onshore, but they nevertheless retain responsibility for home and family. When the woman goes offshore the husband must take sole responsibility for the family during the wife's absence, which means not only that his work-load is increased, but also that he has to maintain the social relations of the family. Often the men have to stay at home much more than they are used to doing, which may result in loneliness and separation from community life. At the same time they usually get more involved with the children, and may find satisfaction in that.

In most cases the woman enters the role of the housewife and takes over the main responsibility for the family when she gets home. Both she and the rest of the family will often feel that her absence has to be compensated for, more than in the case of a male oil commuter, and she will usually become deeply involved in family life throughout her stay at home. Thus, the women workers' period onshore is not 'free time' in the same sense as that of the male oil worker. For a woman offshore worker, almost all the time onshore is work-time: she seldom has any time that is her own, and these women never say that the time onshore drags, as many of the men do.

The family arrangements of the female offshore worker are by no means just a mirror image of those of the male worker. While commuting for a man may be seen as an extension of a traditional pattern of family roles, when a married woman commutes, her work often results in a step towards reversing the sex roles in the family, or at least towards making them more similar. This challenges the intimate relation between family role and gender identity, in much the same way as the relation between work role and gender is challenged when women offshore start working with traditionally male jobs. When this happens in the context of the family, even more deeply rooted feelings of gender differences are aroused, between the spouses, within the family and within the community. It is therefore understandable that the number of married women with children working offshore are few, and are mostly women with a high degree of commitment to their work. They are nevertheless pioneering new ways of combining gender, work and family life.

SOME CONCLUDING REMARKS

The positive effect on the social environment of having women on platforms is considerable. Women workers have not had the disruptive influence on the platform community that was feared by some. Behaviour is strictly controlled on the platform, and having women on board has not changed this. The women themselves are subject to a higher degree of control than are the men, and they also play a part in policing both their own behaviour and that of men.

Instead of creating sexually more loaded surroundings, the relations between men and women often appear almost asexual. This is partly due to the strict social control of those relations, but also to the desire of both men and women to avoid disruptive effects of sexual conflicts. In some cases the presence of women seems to desexualise the atmosphere by curtailing the way in which all male communities tend to objectify women.

To have both sexes on board 'normalises' the atmosphere and this also lessens the differences between onshore and offshore life, which in turn makes the transition between these two periods easier, lessening the degree of psychological change and making commuting somewhat less of a burden. However, the effects women have on the social milieu offshore depends on their number, and on where in the offshore organisation they are found. It may, therefore, be suggested that increasing the number of women working offshore and employing them throughout the platform is important, both for the men and for the women workers themselves.

Women working offshore have also had an impact on the material quality of the work environment. Because they do not have to fulfil male standards of hardiness or strength, women have had little to lose in questioning poor working conditions or unnecessarily heavy work. Some female workers have actually been elected safety delegates and have devoted considerable effort to improving the working environment.

In all probability, women's entrance into offshore work has also contributed to making oil platforms more ordinary work places, where there are now fewer excuses for not bringing the work environment up to standard. Their participation has contributed to a change in the whole thinking about what offshore life ought to be like. The more women obtain access to various positions, and the greater their number, the greater the impact they will have on the organisation of offshore work.

Notes

1. Platforms are referred to as 'integrated' when production, drilling and living quarters are situated on the same installation.
2. Both interviews and field work were conducted in connection with the research project 'Offshore life, family and local community', which has been carried out at the Work Research Institutes, Oslo, since 1981. (See also Chapter 5.)
3. For more inclusive descriptions of platform organisation see Solheim and Hanssen-Bauer, 1983; Holter, 1984; Rogne, Qvale and Ostby, 1982; Qvale, 1985; Heen, *et al.* 1986.
4. The rotation varies. On Statfjord and Frigg the schedule is two weeks offshore, twelve days onshore, twelve days offshore, twenty-four days onshore. The drill crews work sixteen days offshore and have twenty-four onshore. Other contractor groups have different schedules.
5. Most of our data on women in the administration is from the Ekofisk field.
6. For a more general discussion of female office workers see McNally, 1979.
7. For a more comprehensive theory of joking relationships, see Radcliffe-Brown, 1952.
8. Eichenbaum and Orbach (1983) argues that one of the results of the different socialisation of men and women is that men are taught to be dependent upon women and to trust women in giving them support. Women, on the other hand are taught to give support to men, but they seldom get the same emotional support themselves. The result is that men's dependence remains invisible because it is fulfilled, while the female's dependence is more often seen, because it is not fulfilled.
9. The Norwegian oil company Statoil has decided to employ the catering workers themselves on the 'Gullfaks' platforms, rather than contract the work out. This has been a demand of the catering unions for a long time, in order to give the catering workers better job security, and better conditions of work, but it is also seen as a strategy for improving their status on the platform.
10. On drilling rigs catering work may be the starting point for on offshore career which continues into other work, as described in Chapter 3. On production platforms the position is different and catering is set apart from the rest of the work.

3 Women on the Rigs in the Newfoundland Offshore Oil Industry[1]

Dorothy Anger, Gary Cake and Richard Fuchs

INTRODUCTION

There can be little doubt that there are both institutional and interpersonal barriers to female participation in the labour force of the offshore exploration industry. In the family, the school system and the labour market, men and women are led to expect that they will live their lives in predictable ways. The limited presence of women in industrial occupations is therefore but a partial reflection of the patterned expectations and aspirations which individuals come to adopt as their own.

In the last decade there have been both social and economic forces at work which have led individuals to alter their perceptions of what is appropriate in the world of work. Women's increased educational levels; the declining birth rate; the slow growth rates of industrial economies, which create the requirement for multi-income households; and the feminist movement have all contributed to greater numbers of women seeking work and having higher income and status expectations associated with their occupational aspirations. In Newfoundland, where the official unemployment rate hovers at around 20 per cent, women have been affected by these factors and have increased their share of the labour force by more than 10 per cent in the last inter-censal period (1971–81). This increase has occurred in the primary (+4.6 per cent), secondary (+14.7 per cent) and service (+9.8 per cent) sectors of the labour force.

The Province of Newfoundland has been in the throes of the exploration phase of the offshore oil industry since 1966, with the pace of activity increasing since the discovery of the major Hibernia field in 1979. The offshore oil industry is little different from other industrial sectors of the economy with respect to the participation of women. A woman working offshore is the exception rather than the rule both in seeking offshore work and her choice of a career. In this essay we will define the social and economic forces which prevent women from seeking and finding work in the traditionally male-dominated offshore oil exploration industry in Newfoundland. We will, as well, identify the strategies used by women to cope with and overcome these barriers.

The research upon which this paper is based was conducted by the Petroleum Directorate, Government of Newfoundland and Labrador between January and March 1985. It involved semi-structured interviews with women who were working or had recently worked on rigs and supply boats in the offshore sector of the Newfoundland oil exploration industry (including residents of both Newfoundland and Nova Scotia), women who were registered with the provincial government Department of Career Development and Advanced Studies as seeking offshore work, and personnel officers or operations managers of the major oil companies, drilling contractors and supply and service companies operating in the Newfoundland offshore. Table 3.1 provides an overview of the research samples upon which the primary research is based.

Interviews were also conducted with public officials responsible for either promoting women's participation in the oil labour force or providing oil industry related training opportunities. Supplementary additional material was derived from previously unreported interview data from a 1981 survey of the offshore labour force (Fuchs, Cake and Wright, 1983), and from a 1985 household survey undertaken with the Trinity-Placentia Development Association, a

Table 3.1 Research sample (proportions of respondent populations)

Respondent type	*Population*	*Sample size*	*(%)*	*Research method*
Female registrants	84	10	(21)	Telephone survey
Female employees	37	17	(46)	Interview
Personnel officers	57	20	(35)	Interview
Employment coverage	1389	1037	(75)	

voluntary economic planning and development agency in a rural area of Newfoundland where a concrete platform is scheduled for construction.

WOMEN IN THE NEWFOUNDLAND OFFSHORE: A STATISTICAL PROFILE

At the time of our research there were very few women working or seeking work in the offshore oil industry. We were able to identify twenty-three working offshore from January to March 1985 while at the same time there were 1366 men employed in offshore positions. Thus, women comprised only 1.7 per cent of the Newfoundland offshore labour force. While undoubtedly a small proportion, this represents an increase over the situation in 1981 when only one woman was employed offshore. This small number is perhaps not surprising if one considers that relatively few women have presented themselves for consideration as offshore workers. Of the 12 595 individuals who had registered with the provincial government Department of Career Development and Advanced Studies (as at January 1985) only 154 (1.2 per cent) were female and an even smaller number (eighty-four) had indicated an interest or willingness to work in the offshore sector of the industry.

The fact that there are few women seeking offshore positions is confirmed through other sources of information. Attitude surveys conducted in 1981 and 1984 as background reports to the Hibernia Development Impact Statement reported low levels of interest in offshore work by female respondents (1981, 2.5 per cent prefer offshore work; 1984, 0.0 per cent prefer offshore work). However, even among the male respondents, only a small proportion stated a preference for offshore work (1981, 9.7 per cent; 1984, 11.0 per cent). Industry respondents also report that few women are presenting themselves as candidates for offshore positions. In all cases where our industry respondents were able to provide us with data on this issue (50 per cent of industry respondents) women represented less than five percent of applicants seeking offshore work. In one case a firm which employs more than fifty people in an engineering intensive side of the industry had received 300 applications from women for 'traditional' office administrative and clerical positions while only nine applications had been received for 'non-traditional' offshore jobs.

In contrast to the low levels of interest by women in working offshore, the Trinity-Placentia survey showed that a large number of women would be willing to either leave their roles as full-time homemakers (29.7 per cent) or leave their current employment (41.3 per cent) to work in onshore construction related employment associated with the proposed concrete platform yard in that area. The actual types of jobs for which these female respondents have aspirations, however, is unavailable in our data.

While there can be little argument about the low level of interest by women in offshore positions, the simple fact remains that there are *some* women who have the skills, experience, training and interest to work in the offshore. Indeed the educational levels of the women included in our research are considerably higher than those of their male counterparts, as is illustrated in Table 3.2.

Women working in the Newfoundland offshore oil industry are also less likely to be married and tend to be somewhat older than their male counterparts, as is illustrated in Table 3.3.

At this point it may be useful to compare women's participation in the Newfoundland offshore oil industry with that of women in the North Sea. Despite the fact that the Norwegian oil industry has developed to the production phase and employs more than 15 000 in its offshore sector, women have yet to capture many more of the offshore employment opportunities than they have in Newfoundland. In Norway women comprise just under 4 per cent of the offshore

Table 3.2 Comparative male female offshore worker level of educational attainment

Highest level of education attained	*Female[1] registrants*	*(%)*	*Female workers*	*(%)*	*Male[2] workers*	*(%)*
Grade 10 or less	1	(3)	0	(0)	128	(28)
Grade 11–13	3	(9)	1	(6)	143	(31)
Trades vocational school	21	(64)	10	(59)	101	(22)
University						
No diploma	3	(9)	0	(0)	49	(11)
Diploma	5	(15)	6	(35)	36	(8)

Notes: 1. Sample based on all women registered for offshore employment between September 1983 and November 1984 including eleven prior registrants and three Petroleum Technology Students incorporated into telephone survey sample.
2. *Source*: Fuchs, Cake and Wright, 1983.

labour force, while in Newfoundland they represent just under 2 per cent of those working offshore (Table 3.4).

It should also be noted that the higher Norwegian female participation in the offshore labour force is, in part, a product of deliberate policies by Statoil, the Norwegian state-owned oil company, to accelerate the hiring of women into catering positions on production platforms. We have no comparative data on female

Table 3.3 Comparative marital status and age distribution of male and female offshore workers

	Female registrants	*(%)*	*Female workers*	*(%)*	*Male[2] workers*	*(%)*
Marital status[1]						
Single	15	(68)	9	(47)	172	(37)
Married	3	(14)	5	(29)	254	(55)
Separated/divorced	4	(18)	4	(24)	36	(8)
	22	(100)	17	(100)	462	(100)
Age[1]						
18–30	16	(73)	8	(47)	299	(65)
31–40	5	(23)	9	(53)	113	(25)
41+	1	(5)	0	(0)	45	(10)
	22	(101)	17	(100)	457	(100)

Notes: 1. Marital status and age at time of working offshore.
2. *Source*: Fuchs, Cake and Wright, 1983.

Table 3.4 Comparative distribution of women in offshore employment, Newfoundland and Norway

	Total no.	*Female no.*	*Female (%)*
Offshore registrants (Nfld)[1]	12 595	84	0.7
Working offshore (Nfld)[1]	1 389	23	1.7
Working offshore (Norway)[2]	15 340	588	3.8

Notes: 1. As at January 1985.
2. As at August 1983 (Moore and Wybrow, 1984).

participation in the earlier (or indeed current) Norwegian exploration activity, before these limited affirmative action policies were initiated.

Female workers in the Newfoundland offshore oil industry, like their counterparts in Norway, tend to be located in junior catering and other support roles on the rigs and supply vessels. This is despite the fact that, as noted earlier, Newfoundland females working offshore tend to have higher levels of formal education than their co-workers. In Norway, 73 per cent of the women working offshore are in catering positions, while in Newfoundland 48 per cent are in these occupational categories. In Newfoundland there are no women working in the 'hard handed' drilling jobs of the offshore exploration industry. Table 3.5 illustrates the comparative occupational distribution of males and females in the Newfoundland offshore.

Table 3.5 Comparative percentage distribution of male and female offshore workers by employment categories

Employment category [1]	*Males*[2]	*Females*	*Female respondents*
Manager consultant	0	0	6
Drilling personnel (senior)	6	0	0
Drilling personnel (junior)	21	0	0
Marine personnel (senior)	23	9	6
Marine personnel (junior)	17	0	0
Service personnel (senior)	10	43	18
Service personnel (junior)	11	48	71
Other	3	0	0
	100	100	101

Notes: 1. Purpose designed occupational classifications, Department of Rural, Agricultural and Northern Development, Research and Analysis Division, 1981.
2. *Source*: Fuchs, Cake and Wright, 1983.

THE HIRING PROCESS

With an unemployment rate exceeding 20 per cent, most oil industry employers in Newfoundland find it unnecessary to use active recruitment policies to find workers. Instead, most positions are filled either through word of mouth or from the numerous applications which are already on file at the company office.

An industry respondent described the employee selection process this way:

> There's a continuous stream coming through the door. It's word of mouth. You don't have to spend a penny on advertising, that's for sure. We only advertised once just to let them [potential applicants] know we were here.

The large pool of labour and the subsequent use of passive recruitment have several effects on the participation of women in the offshore labour force. First, the word-of-mouth or 'buddy' system of filling vacancies is more likely to benefit males in a situation where 98 per cent of the labour force are men. The chances are that an interested woman will not learn about prospective job opportunities through this process. Secondly, the woman who has shore-based skills and experience which qualify her for an offshore position is unlikely to associate her skills with an oil rig because the companies have no need to advertise locally for qualified workers for most positions.

It is interesting to note that most of the women working offshore were already highly qualified for their positions and were, in fact, beneficiaries themselves of the direct visit to the company office or the word-of-mouth system in securing their jobs. A study of an onshore concrete platform construction site in Norway showed that the women in the catering staff relied on the 'buddy' system in order to obtain a job and a congenial work environment. Leira (1978) writes:

> Approximately one-half of the women interviewed were recruited informally by friends, relatives and neighbours. The work at the Condeep site was, for many, not just a choice of work place but also work companions.
>
> (Leira, 1978, p.33)

The impression we gleaned from our industry interviews was that many of the companies want to hire more women for what might be termed 'traditional' occupational roles offshore. Few women, as we have already noted, are presenting themselves for prospective employment, in part because of the excessive labour supply and the subsequent recruitment policies which are being used.

We were also interested to learn that the influence of government agencies in securing the participation of women in the offshore was considered to be negligible by both industry and female offshore workers. Indeed, a 'double-bind' situation exists whereby those women with the necessary skills to secure offshore work feel no need to rely on government agencies for assistance. They go directly to the company office. On the other hand, those women who are least qualified seem to rely upon and be referred by government agencies more frequently. Sometimes this results in, or reinforces, industry skepticism about the usefulness of public intervention in the hiring process. This is reflected in the following comment by an industry respondent: 'The local (provincial government) office sends in welfare cases. That's a waste of money.'

The influence of government agencies in promoting women's participation in the offshore labour force is nevertheless real. However, their major function is an indirect one, in that they place pressure on the companies to hire women. It happens that this results in the more qualifield and self-directed women being hired rather than those who are referred by government agencies. This is reflected in the following statement by a female respondent:

> If the operator is told [to hire women] then it filters down through the companies, but it may not get out to the rigs. So they were looking for a female. That's why I got it [the job]. I was the only one that applied – the only woman – *and I was qualified*. It was a windfall for them and for me.

It, therefore, appears that oil industry firms are willing and, in fact, interested in hiring women for some types of jobs offshore. They are constrained in this by the absence of sufficient numbers of qualified applicants.

WORKING OFFSHORE

If we accept that few women are interested in working offshore, and that the Grand Banks of Newfoundland is a generally inhospitable work environment for men and women alike, the question 'why are any women working offshore?' presents itself. A second and related question is, 'to what extent is their work life affected by being female in an almost exclusively male world?' In order to answer these questions, and to examine more closely the complexities of the offshore work and social environment, we will look at the consonance and dissonance in values, motivations and experiences of female offshore workers compared to those of their male co-workers.

Women working offshore, and their motivations for doing so, are as diverse as those of their male counterparts. Some agreed with one women's statement that, 'I'm a sailor, I just want to be on the sea.' For others, the work schedule and the intensity of work and social relations make the regularity of onshore work difficult to contemplate, 'I couldn't work nine to five, day in and day out, again. I don't know what I'd do if I had to leave that rig.' Others have specific goals and timetables which do not include a long-term commitment to the offshore; in the words of one, 'No one *wants* to be offshore. Everyone is doing it to get the money that will allow them to do what they really want to do.' A similar range of motivations and attitudes towards their work has been documented among male offshore workers (Fuchs, Cake and Wright, 1983).

Women seeking offshore employment and those working offshore mention the money and the time off as primary motivations. One might speculate that, especially for women in catering positions, these are considerable advantages, given that waitresses, chambermaids and cooks make only the minimum wage,[2] or slightly above, in onshore positions. A woman who worked in the galley said, 'Sometimes you get fed up with it out there, but then you think, "think about how much you'd be making doing this onshore."'

All of the women working offshore reported continually weighing advantages and disadvantages of their work. The money, time off, and career advancement may motivate them to remain in the industry even if that has a cost to their personal lives. However, in order to make a long-term commitment to working offshore, career advancement may be the single most important factor. And, in the positions held by most of the women working offshore, the opportunity for advancement is very limited. The effect of that on women's employment plans was very evident in our interviews. Most women said something similar to the following:

> I'd like to make a career of it. But not if all I can do is stewarding. And, where I'm working, it's once a steward, always a steward. I like it alright, but I don't want to do only that for the rest of my life.[3]

The nature of the work requires relatively long periods of time away from home and intensive contact with a relatively small group of co-workers. As the literature on family adaptation to offshore work suggests, this pattern may place stress on onshore family and social relationships. Although we have not explored this area extensively, some themes became evident during our discussions with female offshore workers. Most of the women who have a strong career commitment to the offshore are either single or have partners who also work in the oil industry or in some comparable occupation. However, many who are single foresee problems in establishing or maintaining personal relationships while working offshore. Because of the traditional expectations and roles assigned to women, we may suspect that the conflicts presented by work which requires extensive absence from home may be even greater for them than for men. While the women to whom we talked have successfully reconciled their work and their personal lives, they acknowledge that there are inherent difficulties. While it is difficult to generalise from individual plans and motivations, we may say that those who have a strong commitment to a defined career path or to the offshore community will give the industry priority over their personal lives; a choice that men do not have to face in such absolute terms.

Plans to remain in the offshore oil industry do appear to be correlated with the degree of adherence to what we might call the ideology of the industry. Interviews with local oil workers (Fuchs, Cake and Wright, 1983) suggest that Newfoundlanders, both male and female, are no less susceptible than others to the lure of money, excitement and promotion evinced in oilpatch imagery. One woman said:

> You can work thirty years in an office or teaching, and end up making forty thousand dollars a year, retire on maybe a half decent pension, and you've done really well. But you can work in oil for twenty or thirty years, and start with nothing and end up making one hundred and forty thousand a year. You can't do that anywhere else.

But can she realistically ever expect to move close to that realm? She was a cook. Cooks earn $25 000 or $30 000 a year, and may move into chief steward positions. These are higher paying and more responsible positions, but both male and female workers have questioned whether women would be able to do even this job, because it involves authority over others. A woman said: 'The men won't take orders from a woman, no matter how well you know your job.' A man interviewed for *The Steel Island* (Fuchs, Cake and Wright, 1983) confirmed her opinion, saying about women working in the galley, 'You're going to have to have a male chief steward over them to deal with the complaints.'

The offshore jobs presently held by women are either at the lowest end of the salary scale or they are in relatively well-paying but specialised services. Leaving aside questions of male–female interaction and authority systems, the promotional ladder for the range of occupations held by women is truncated. Galley stewards may move to cook, cooks may move to chief steward. Sample catchers may move to mud-loggers, and to the limited range of geological positions. Medics and technical support services, such as radio operator and weather observer, are occupationally isolated with no direct steps above them. Therefore, the positions in which women are now working are relatively limited in real terms and in terms of occupational mobility. In addition, questions of women's abilities to 'do the job' or to be in positions of authority have thus far prevented women from moving outside the service sector.

There is a well-defined and traditional employment mobility pattern in oil exploration. Stewards may move to roustabout – a move from inside to outside. Roustabouts (general labourers) move to roughneck – the first step on the drill-floor – and up many steps to the top position attainable through drill floor promotion, toolpusher. In theory, each step upward for one worker frees the path for another coming behind. It also encourages belief in the oilpatch ideology, that there is the potential for fast advancement, big money, and authority, in which rewards are not based on paper qualifications, but on individual ability and a willingness to work hard.

While some women subscribe to this belief, most know that for them it is unattainable. Their reactions are varied: some plan to remain offshore for a limited time; others obtain the education which may allow them to skip a few offshore steps; and others have confronted the system on the rig and tried to enter 'outside' jobs.

Their experiences suggest that, while the existence of overt discouragement towards women has lessened since they started working offshore, discouragement still exists with varying degrees of subtlety. This is especially true when women attempt to go outside the service occupations.

Thus far, the positions which we have discussed are traditionally female jobs, concerned with 'reproductive' tasks such as cooking, cleaning and nursing. All that makes them 'non-traditional' is their location. It is also interesting to look at attitudes toward women in non-traditional jobs in the most extreme form on an oil rig – jobs on the drill floor. This serves as the most effective illustration of the influence of social and cultural factors affecting women's position.

One industry respondent said about women on the drill floor, 'I can't imagine any woman in her right mind putting herself in that environment. It's just flat ass tough work.' The offshore work-place is seen as a hostile and rugged environment by the general public and industry insiders alike. As a separate and total institution, the offshore oil rig has its own rules, its own value systems, its own culture, and the drill floor epitomises that image and its reality. Especially on the drill floor, the formal and informal rules and authority system are central to doing the job and not jeopardizing safety. Successful rig workers are those who learn rules quickly and adopt them as their own – and this is most sharply defined on the drill floor. Interviews with industry representatives and the women themselves provided an indication of the impediment posed by the culture of a drilling rig to those considered incapable of sharing its rules, especially women. As one industry man said:

> You can't shit on a woman the same way you can a man. You can put that in whatever sociological jargon you want to. It's authority by intimidation. The tension of the drill floor is intense.

While drilling, and hence the culture of the drill floor, pervades all areas of a rig, the intensity of action and authority is somewhat lessened away from the drill floor. Relaxation and leisure time occurs inside, where women work, and where their presence is acceptable and welcomed. However, there seems to be unanimity that women's presence on the drill floor would be unwelcome.

Among those to whom we talked, no woman who has worked offshore expressed any desire to work on the drill floor. However, their reasons included both the severity of the work and the lack of acceptance by the men and the industry. Most thought that some women would be able to do the work, but all agreed that, as one said, 'It'd have to be a pretty brazen woman to do it.' Another explained this further, 'I can't see the men tolerating a woman. They'd quit first. I don't know why a woman would want to do it. It's too hard a job and who'd want to put up with that from the men?' If women are viewed as unlikely and unwelcome intruders into the 'business end' of offshore exploration, it should come as little surprise that they do not include opportunities on the drill floor within their range of employment aspirations.

Deck work encompasses a variety of less specialised and pressured jobs, such as maintenance, roustabouts, etc. Women do include such work in their range of employment possibilities, but their view has not been shared by industry officials to the point of hiring women for these positions. At least one industry representative held women's own reluctance to be the problem:

> If they're not willing to approach us, I'm not going to dig them out of their holes. It's more that they have been trained away from these jobs – that's more important than physical strength. You don't have to be Charles Atlas but you do have to be physically fit.

However, the experience of women who have demonstrated that they *are* interested and able to do deck work is that they are passed over. As one woman said, 'You've told everybody you want to go outside, and it still ends up with six guys ahead of you. I even stopped asking. It's too discouraging.' None of the women interviewed suggested that they should get priority over men who were also waiting for jobs, but they did ask that they get the same treatment. That means getting the next job available if they have been waiting the longest.

Aside from drill floor jobs, exclusion from deck positions means that many of the mechanical and trades positions are also closed to women, because apprentices are usually recruited from the deck and the general knowledge gained there is necessary to supplement other forms of onshore training. With the drill floor being totally off limits and the deck almost as completely closed, there is little upward mobility possible for a woman on a rig. Given the isolation and rigour of the offshore environment, it is perhaps surprising that as many women have remained working offshore when there is such limited opportunity for advancement.

But this is painting a misleadingly negative picture of what the women think about their work. Those to whom we talked, while recognising the stress entailed by the job itself and in being women in a masculine environment, liked what they were doing. None of the former workers interviewed had left the offshore because they could not tolerate the work or the environment, although the degree of their affection for it varied. Their understanding of the rig culture and their ways of adapting to it explain the dynamics of social relations offshore and provide a model for how women can work in such an environment.

THE POLITICS OF SOCIAL RELATIONS

An offshore rig has been called a 'total institution' (Fuchs, 1983; Holter, 1984). Its culture is one of cross-cutting systems of authority and community, enveloped by isolation from home on the one hand and cheek-by-jowl existence with co-workers on the other. Fuchs described it as follows:

> The offshore oil rig is a total institution. The status and roles which a person occupies in his home community bear no relationship to his position in the offshore oil industry. The frequent readjustment from a rural community where a person holds a prescribed status, derived from family relationships and particularistic standards of acceptance within the home, to the offshore oil rig, where a person is confronted by a pressurized, achievement based system of reward and stratification can be difficult.
>
> (Fuchs, 1983, pp.21–22)

One woman described the politics of the rig thus:

> It's as if you went home at night and slept next door to your boss. You can never get away from the fact that you're working. People aren't human beings, they're parts of companies.

The women to whom we spoke are very conscious of the existence of a well-defined status hierarchy, and of the confusions presented by a situation in which hierarchy overlaps with friendship. As a woman said:

> There's lots of status things that you have to know about – chains of command that you can't cross. Oh, you talk to everyone, but you have to remember who they are and where they fit in.

Authority, and prestige, radiate outward from the centre, which is the drilling operation.

However, along with a hierarchical authority system is a sense of community developed among eighty disparate personalities. One woman said, 'We're close. I think some of that may be due to having women out there, but even amongst themselves, they get along really well. I start to miss them when I'm onshore.' For men and women, the intensity of contact and the work itself serves to create strong bonds between workers. However, it also appears that a social etiquette exists which prohibits emotional outbursts. There is a constant tension between camaraderie and maintenance of social distance, which keeps potential conflicts at bay, but also places considerable stress on the individual. All the women interviewed said something similar to the words of one:

> You have to live with the same people for twenty-one days. You can't start fights, and you can't get really close to anyone. You just have to get along with everyone equally well. Oh, you can be best buddies, but there still is always some distance, some privacy kept.

In this total institution, emotional bonds and 'buddies' do exist, but interpersonal relationships are tempered by the knowledge that no matter how much one likes or dislikes another, all must abide by an etiquette of civility.

Women come into this complex offshore environment not only as employees and individuals who must find their own niche, but have to do so as rather conspicuous novelties amid a sea of men. A woman described going to the messroom:

> The place is blocked – with men. They watch you from the minute you walk in, they watch what you eat, where you sit, how long you stay, how many cups of coffee you have and how you drink it. It's like *you're* the entertainment.

A male rig worker said the presence of women was good for the morale of the rig, as he said, 'It's somebody different to look at. It makes you feel better just to known they're there.'

The role of friend and confidante is one well understood by the women working offshore. Most of the women see their presence as important in recreating a sense of home. As one said:

> The guys can talk to me about things that maybe they wouldn't talk about to the other guys. It's not a masculine thing to say,

'Suzy took her first step today.' But they know they can tell me and I'll be excited with them.

However, acting as social facilitators also places greater stress on women. One woman said, 'I'd like to see twenty more women on my rig to take some of the pressure off me. A woman to talk to is in big demand.' In addition to relieving the social pressure on each one, more women would also give each other support which many feel they now do not have. One woman said, 'It's a harsh environment, and the guys have each other, but the women don't have anyone.'

Other forms of work-related stress also develop for women due to their small numbers. Although their social obligations are considerable once they have become accepted by the men as fellow workers, the period prior to acceptance may be longer and more difficult than for new male workers. Because of the rarity of female presence, sexual tensions, and some antagonism towards 'women doing men's jobs', the men may be reticent in befriending new female workers. If a woman feels unwelcome and ostracised, as some reported, the adjustment to an alien and harsh environment will be more difficult.

Women must also be able to deal with sexual joking and innuendo. One woman gave us some advice, 'If you're gonna go offshore, just remember to bring a big, baggy sweater with you. That's a girl's uniform out there. You don't need to be singled out more than you already are.' However, perhaps surprisingly, the level of sexual harrassment reported by the women is probably lower than on the average construction site. This is, in all likelihood, due to the maintenace of equilibrium between camaraderie and social distance in order to co-exist for extended periods of time with the same people. The attitude of the women also prevents 'carrying on' from becoming harrassment or scandal. The consensus of the women is that the best way to behave is to be 'one of the boys' to some extent, but not to emulate 'the boys' to the point of not being a woman. One said, 'The best way is just to be nondescript.'

Trying to blend into the crowd is important for reasons other than avoiding unwanted advances. The sense of community and the functioning of information networks are, as everywhere, based on similarity and shared experiences. Appearance emphasising 'otherness' may be enough to exclude women from vital support networks. One woman said:

> You'll get along fine with the guys if you're prepared to be one of the boys. If you want to be a 'femme fatale', they'll like it, but you're going to be kept in the dark about what's going on.

The need to 'blend in' extends to the women's attitudes toward their work. They agreed that their work demands should be no less and no more than the men's. They do feel that the social demands on them are greater, but are prepared to accept that. However, many feel that their performance on the job is more closely monitored than that of men. This is perhaps due in part to their higher profile, but also to an underlying opinion on the part of some men that women cannot adequately do the work. Feeling that they have to work to higher standards than their male counterparts discourages and angers many of the women. One said:

> To have to do that extra, to work that much harder just to keep your job is such a drain. A lot of men don't think that women are being treated any differently, but *aware* men know that's not true.

This makes the warnings against 'expecting special treatment', which most reported receiving at some point, even harder for them to accept graciously.

Their desire *not* to be treated differently extends to the area of affirmative action programmes. One woman stated this opinion forcefully, 'What I don't want to see is people like you, in positions of power, pushing to get women out there, whether they're suitable or not. That is only going to make it hard for women out there now.' They recognise that the nature of the work means that some people – male and female – cannot survive offshore. However, they also recognise that the significance of a woman not being able to do it is much greater than if it is a man. They believe that the actions of one woman reflect on others and, while they do not like it, they have a vested interest in seeing qualified and competent women working offshore. They recognise the danger of 'getting numbers', and want to be regarded as workers rather than as 'target group members'. However, they also believe that they have a responsibility towards other women. As one said, 'I didn't go out there to be a pioneer, I wanted a job. But I know that is what I am, and how I, and other women turn out, will affect how many more they will hire.'

CONCLUSIONS

In presenting a statistical and ethnographic overview of women in the Newfoundland offshore exploration industry, there are two concluding points which are worthy of mention. First, it is important to bear in mind that women working offshore identify themselves, first and foremost, as workers. The salience of their occupational identification is unmistakeable and, as has already been discussed, our respondents reported using covering and concealment strategies to remove the issue of gender from their interpersonal relations in the work place much the same as minority interests in other mixed contact settings. Their skeptical attitude towards affirmative action, their tendency to set very high work standards for themselves, and their willingness to adopt the work place ideology of individualism are but a few of the manifestations of the women's identification as workers.

In placing the greatest emphasis on their work role, women in the offshore oil industry are little different from their male counterparts or from women who work in other isolated and male-dominated industrial settings as the following reference to mining in the Yukon illustrates:

> Most important, however, is that to a considerable extent their [women's] work roles and social relations are formed by aspects of mine work unrelated to gender. Simply put, Cyprus Anvil's women production workers most often regard themselves, and are regarded by others, as mine workers first and foremost: as drillers, pit labourers, equipment operators and so on, rather than as 'women in men's jobs'.
>
> (Martin, 1984, p.24)

A difference, however, between the Cyprus Anvil miners and female offshore workers is the fact that the latter are confined to occupations which are largely offshore extensions of 'traditional' shore-based occupations. While they see themselves as workers and go to considerable lengths to present that role in their occupational relations, they are confronted with a raft of expectations to be nurturant, sympathetic and congenial by many of their male co-workers. These expectations present women offshore workers with a role conflict. They are perceived in ways which are at variance with their own self-image.

Secondly, we believe that this role conflict can contribute to the levels of stress which are already attendant with work in an isolated total institution on the North Atlantic. A recent report published by Mobil Exploration Norway Incorporated suggests the same conclusion when it describes the results of a health and safety study conducted by social scientists on the Statfjord A Platform as follows:

> However, smaller groups exhibit several signs of an unsatisfactory health condition, such as frequent and troublesome symptoms, use of medication, sleep disturbances, etc. Among these are more women than men, they are poorly educated, have no experience at sea . . .
> These groups also experience several aspects of working offshore as unsatisfactory or stressful.
> (Mobil Exploration Norway Inc., n.d., p.12).

Despite the institutional and interpersonal barriers women experience in the offshore work place, it is our impression that they have adapted well to the work environment and their presence is welcomed, at least in 'traditional' roles, by their male counterparts. We are confident that, where they are presented with promotional opportunities for 'outside' jobs, they will be able to perform well and, where their presence in the offshore becomes perceived as being normative, greater numbers of qualified women will present themselves as candidates for offshore work.

Notes

1. This paper formed the basis of a presentation by Irene Baird (Assistant Deputy Minister, Policy and Planning, Newfoundland and Labrador Petroleum Directorate) at the International Conference on Women and Offshore Oil, St John's, Newfoundland, September 1985.
2. In Canada, legislation requiring payment of a minimum hourly wage to all workers was introduced in 1935. Similar legislation was enacted in Newfoundland in 1953. At present, the Newfoundland minimum wage is $4 per hour for all workers, except domestics, who receive $2.75 per hour. The Canadian rate is $3.50 per hour, and $3.25 for workers under seventeen years of age.
3. Similarly, the study of the impacts of offshore employment on Newfoundland families showed the highest levels of dissatisfaction among male rig workers at the lower end of the employment hierarchy (Community Resource Services (1984) Ltd, 1986, p.43).

Part II

Women, Family and Offshore Oil

Introduction

The literature on the experiences of families in which a spouse is intermittently absent has been dominated by studies of military families, particularly of Americans during the period of the Vietnam War (for an extensive bibliographical review see Community Resource Services (1984) Ltd, 1986). Many of these studies make implicit assumptions about the nature of family life and relationships, most favouring the male breadwinner and female and child dependants model, with its accompanying division of labour, paid and unpaid, between husband and wife. This in turn has led to the problematising of spousal intermittent absence. Indeed, older studies tended to view the wife's responses to the periodic absence of husbands as pathological and also tended to blame them for failing 'to adjust' (e.g. MacIntosh, 1968). Such studies focused on the individual determinants of behaviour and on individually oriented solutions; most advocated some form of psychotherapy for the wife.

The essays which follow do not deny that offshore oil work poses problems for the families of offshore workers, the vast majority of whom, as we have seen, are male. (Of those workers who are women, most are single.) Indeed, the study from which Clark and Taylor's chapter is derived began by hypothesising the widespread existence of something called 'intermittent husband syndrome', characterised by symptoms of anxiety, depression and sexual difficulties. The existence of the syndrome in the wives of offshore workers was widely reported by general practitioners, health visitors and social workers in Aberdeen; however, in the event the study (of some 200 offshore wives) found a rather low (10 per cent) prevalence of 'caseness' among them. While this study continued the practice of assuming the existence of pathology as the result of intermittent spousal absence and of focusing on the behaviour and feelings of wives, it no longer sought to apportion blame for the existence of family difficulties and also attempted to consider a wide range of socio-structural as well as individual variables.

The rotational work patterns associated with offshore employment which result in the continual absence and continual presence of the worker for two, three or four week periods do pose problems for family life, the variety and intensity of which vary according to a wide range of factors, including the expectations the partners have of marriage and the resources they can bring to bear in negotiating their situation. The latter may depend as much on educational achievement, geographical location and kin and neighbour networks as on personal qualities. Furthermore, the process of negotiation may be harmonious or conflict ridden, and more or less favourable to one partner. In the vast majority of cases, the male partner goes offshore to a work culture that both wholly excludes and denies family life. The Newfoundland government's report on the experience of rigworkers described the hierarchy aboard the rig as 'para-military' (Fuchs *et al.*, 1983). Offshore oil work, as Holter (1982) has remarked, offers little beyond the rewards of higher than average pay and what is often perceived as considerable leisure time. The reintegration of the worker into the world of home and family might therefore be expected to pose problems. In this analysis, it is the nature of offshore and the rig/platform culture that should be regarded as the source of possible family problems, not the family itself. The essays that follow tend to support Holter's observation based on the Norwegian experience: 'Many of the home problems of the offshore workers are in reality work problems'. If this is so, then an individually oriented therapy solution cannot be expected to suffice.

The nature of the effects of offshore work on family life are difficult to assess. The Aberdonian study of oil wives began by delineating three main types of response to husband absence (Taylor *et al.*, 1985; Morrice *et al.*, 1985). The first was characterised by loneliness and depression, the second by a determination on the part of the wife to carry on in much the same way, whether the husband was absent or present, and the third by feelings of resentment, both while the husband was absent and present. However, as the study proceeded, it was found that the reactions of most wives were much more complex. The essays that follow show that the reactions of most wives of offshore workers are profoundly ambivalent. Furthermore, when the reactions of husbands and the interactions between husbands and wives and parents and children are also considered (as they are in both the chapters on Norway and Canada), then the picture becomes additionally complicated. Interestingly, the contributors to this section have often ended up interpreting quite similar evidence regarding the behaviour of husbands and wives very differently, reflecting in part their different approaches, and in part the variations between their perceptions and the reality of patterns of family life in the three countries.

All wives of offshore workers must cope with the cycle of continual absence and presence of their husbands. During the period of absence they are obliged to shoulder all family responsibilities, which may require them to exercise a greater degree of independence than they consider to be either appropriate or desirable. Alternatively, some women may experience difficulty in reintegrating their husbands into their lives during their periods onshore. The nature of the adjustments husbands and wives must make depends in large part on their expectations of marriage. As Harriet Gross (1980) commented in her study of dual career couples who lived apart, both the husbands and wives could not but take traditional marriage as the starting point from which to assess the advantages and disadvantages of their situation. The traditional model of monogamous, heterosexual marriage provided the natural frame of reference for these couples and for the families of offshore workers. In all cases, the rotational work pattern requires some renegotiation on the part of both partners and the capacity for achieving this would seem to depend on social and economic circumstances, personality and the nature of the power relationship between husband and wife.

The data from Newfoundland and Scotland suggest that the majority of families shared the aim of using the higher levels of income available from offshore work to fund family based 'projects', like buying a better house. In both studies a high percentage (48 per cent in Newfoundland, 37 per cent in Aberdeen) of men had previous experience of work that took them away from home. This, together with the fairly low percentage of wives who reacted negatively to the idea of their husbands continuing to work in the industry (19 per cent in Newfoundland did not want their husbands offshore in three years, and 29 per cent in Aberdeen said no to work offshore in five years time), despite the tensions and difficulties they articulated, suggests a considerable mutual commitment to the industry for the sake of the material rewards it delivers. Many of the wives interviewed praised their husbands' efforts to make better provision for their families and in particular for their children's futures. This evidence suggests that a majority of families adhered to the idea that the male should provide and that the primary role of the wife was to maintain home and family.

The Norwegian data also indicate that husband and wife share a common purpose in pursuing offshore work; however, the majority of Norwegian men interviewed in the course of the Norwegian Work Institutes' research project on offshore commuting cite the long periods onshore rather than the financial rewards as the major reason for working in the industry. This is interpreted by Solheim (Chapter 5) as indicative of their commitment to a more companionate marriage and to more involvement in domestic life, especially in parenting. Such a commitment is, of course, not incompatible with a more or less traditional division of labour in terms of paid and unpaid work between spouses. If it is indeed the case that Norwegian men do have greater expectations of participating in domestic life, while Scottish and Newfoundland husbands and wives have more traditional ideas of sexual divisions, offshore work patterns will paradoxically pose problems for the achievement of both ideals. Furthermore, the problems may be seen to be experienced differentially by husbands and wives, with the greater burden falling on the wives. Whether they have more traditional or more companionate ideas about the organisation of married life, wives must face repeated periods of husband absence and the persistent problem of transition from continued absence to continued presence.

The essays that follow show the range of issues and the nature of the conflicts that women experience in dealing with an intermittent husband. During the period of absence, they must act more autonomously in order to survive. In terms of the material welfare of the household, husband and wife must in all probability negotiate the extent to which the wife assumes responsibility for finances and domestic maintenance. The now extensive literature on the distribution of resources within the family (Pahl, 1980, 1985; Land, 1983) draws attention to the large numbers of women who still do not know what their husbands earn (49 per cent of married women in Britain in 1974) and to the wide variety of patterns of money management within marriage. This research indicates that the lower the husband's wage, the more likely it is that the wife will manage the whole of the family income. In better off families, wives may bear responsibility for daily purchases, while husbands pay monthly or quarterly bills. Such patterns are closely related to the reality of female economic dependency. Newfoundland and Aberdeen data show considerable variation in the ways that financial responsibilities are negotiated, with a majority of wives taking over while the husband is away, but passing back overall responsibility when he returns. This may be seen as a compromise designed to accommodate traditional views about sex roles, but one that is likely to produce tensions, especially in a situation where money provides the major incentive for staying in the industry.

Women must also deal with the emotional pressures of parenting alone for prolonged periods and of dealing with their own feelings of loneliness and isolation. All the studies show that children often provide considerable comfort as well as extra responsibility, and that while the wives of offshore workers are additionally reliant on neighbours, kin and friends for help in coping with crises (especially when a child is sick) these networks can be sources of rivalry and jealousies as well as support. Wives may well find paid employment a source of satisfaction during the husbands' absence. But in all three countries the labour participation rates of offshore workers' wives are significantly lower than the national averages for married women, reflecting in part the many structural factors that shape women's work opportunities (particularly stage in the family lifecycle, the availability of child care, and the nature of the local labour market), and in part the difficulty women have in maintaining such external commitments during the husband's period onshore, when the wife feels pulled between loyalty and affection for her job and for her husband. Thus the patterns of offshore work may be seen to exert a powerful influence on the way wives structure their lives and to elicit additional domestic labour from the wife as a necessary adjunct to the husband's absence offshore.

From the point of view of the male offshore worker, whose views were surveyed in the Norwegian and Canadian studies, it seems that male dependence on the family is increased. Men see the family as 'a haven' to which they return from the rigours of offshore work and the harshness of the offshore work culture. As Solheim (1983) has remarked, this tends to create 'an overload of expectations in the home context', which as Clark's chapter in this volume shows, can result in problems at the time of reunions. Men rely on the family 'being there' and providing a place to which they give vent to their work frustrations, either in words or actions, by sleeping, throwing themselves into home based projects, or drinking. The Newfoundland study concludes that it is in fact women who bear the burden of the 'double adjustment' to the rotational work pattern and to the needs and moods of the male worker.

Thus the research reported in the chapters that follow provides some explicit support for David Morgan's (1975) proposition that families may share harmony of purpose or 'projects', the pursuit of which may be none the less experienced differentially by different family members. In the case of offshore oil families the burden of adjustment tends to fall disproportionately on the wife. The Aberdeen study attempted to construct a profile of wives 'at risk' as a result of onshore commuting. The study concluded that wives who were younger, more recently married, with small children, no experience of husband absence and with a personality characteristic identified as 'over-dependency', were more likely to experience difficulties in respect to health, mood and behaviour changes, and greater conflict with their husbands. The Aberdeen study did not include men, which resulted in women's experiences and reactions being identified as the source of both the problem and the solution.

In contrast, the Canadian and Norwegian studies locate the source of both the problem and the solution in the nature of offshore oil work itself. The Norwegian study suggests that those husbands and wives with more traditional expectations of marriage and sexual divisions may experience fewer tensions as a couple, it being more easy for the husband to move in and out of the family without any great readjustment of roles and tasks. A similar suggestion is made by Gramling (1985) as a result of his work on the social impacts of offshore oil development in Louisiana. Following the theoretical work of Berger and Kellner (1964), he suggests that more 'traditional' families will be more embedded within their communities with the roles of husband and wife being more subject to external definition. More 'modern' families (committed to 'being together' in Solheim's phrase) and more dependent in Gramling's view on the identification provided for them by the family, will experience greater strain as a result of offshore oil commuting. However, if the total experience of offshore work is considered from the woman's perspective, it seems that while all the women who contributed to the studies reported in this volume can be said to be 'coping' with the situation (albeit with varying degrees of satisfaction), all have had to make 'adjustments', no matter what their ideas about marriage and the division of labour between husband and wife. While the nature of these adjustments varies considerably, it is not possible to conclude that one set of difficulties was intrinsically more problematic than another. For example, the Newfoundland material documents the often agonising loneliness experienced by some women, but also the difficulties of wives who want to work but experience tensions in so doing while their husbands are at home.

On the face of it, offshore work offers both the opportunity for the development of greater female independence and greater male involvement in the home. But in practice, the pattern of continual absence and presence militates against both. Men must move between two environments which are at present hopelessly estranged in terms of their respective cultures and it falls to women to 'adjust'.

4 Partings and Reunions: Marriage and Offshore Employment in the British North Sea[1]

David Clark and Rex Taylor

An earlier generation of sociologists would have looked at the oil industry in the context of a body of work concerned with 'extreme occupations'. Much of that material was predicated upon the notion of a radical separation of home and work, apparently visible in the lives of fishermen (Tunstall, 1962), lorry drivers (Hollowell, 1968), miners (Dennis, Henriques and Slaughter, 1969) and others. This separation allowed the *work-place* to become the focus of attention, reducing family, kinship and community to the level of dependent variables. As we have already seen (pp.12–19) such a distinction often serves to mask the complex interrelationships between work and family and is likely to obscure important gender differences with shape experiences in the two arenas. A growing body of writing has therefore begun to examine the work/family nexus and the concept of *incorporation* (Callan and Ardener, 1984) has been especially helpful in revealing the variety of ways in which women may be co-opted as unpaid collaborators in their husbands' paid employment. Whilst occupations such as the ministry, general practice or those in which the worker is self-employed provide obvious examples of this phenomenon (Finch, 1983), some of the very jobs which were once classified as 'extreme' in character also merit attention. Indeed the paradox of these occupations is the degree and special nature of their impact upon domestic relations. Detailed examination of such an occupation therefore constitutes an important case study, revealing trends which are equally salient, but less visible in more familiar occupations.

It is, however, important to deconstruct what we mean when we talk about the work/family nexus. The relationship is often depicted at a fairly high level of generality; moreover it invites definitional questions about the terms in use. We shall want here to concentrate on a particular dimension of the nexus and to look at the relationship between occupation and marriage. This is important because our data reveal a number of significant implications for the wives of offshore oil workers, who it may be argued are confronted by issues not shared by the wives of men with similar jobs onshore, or perhaps by the wives of men who work away from home in other industries. It is therefore the highly *particular* combination of employment in the North Sea oil industry and its effects upon the marriage relationship that is our concern.

In this chapter we shall be looking at various aspects of this issue as revealed by a Scottish study[2] of women married to or living with men employed in the offshore oil industry. We shall pay particular attention to the implications of offshore employment for the marriage relationship, especially as this is influenced by the women's self-image and we shall also look at its influence on parent-child relationships. Our aim will be to understand how the regular pattern or partings and reunions experienced by the oil family may affect individual members in different ways, as well as having an influence upon the family's lifestyle more generally.

As we shall indicate below the initial focus of our research problem was a clinical one, concerned especially with women's reactions to their husbands' employment offshore. Nevertheless, from an early stage in our enquiry, we gained insights into a broader set of factors which contributed to that experience. These reveal some of the interactions between work, marriage, and family life and demonstrate the need for an inclusive approach to studying them. For example, having obtained funding for our projet in 1981, we set about examining various ways in which a sample might be generated. We made a variety of attempts to secure the cooperation of the oil industry in this, both by approaching individual companies, as well as more representative bodies, such as the United Kingdom Offshore Operators Association. Our assumption in this was that the industry would see the value of a study which examined some of the consequences of its patterns and modes of employment. This was of course a naive view. It turned out that *whatever* such consequences might be (and we emphasised the merits of a 'scientific' study of these, rather than journalistic inference) neither the industry as a whole, nor any individual company was prepared to assist in a project which might increase the level of public scrutiny. This defensiveness was apparently based upon certain views about the work/family relationship. We discovered, for example, that some employers only made formal communication with their employees at the work-place and avoided writing to them at home. Moreover, if there were any domestic consequences resulting from offshore employment these were to be regarded as an unfortunate but inevitable aspect of an occupation which demanded a great deal of its workers, but whose financial rewards were seen as adequate recompense. As one personnel officer put it at a meeting we attended: 'we employ men not families'. It might be argued that the employment 'contract' is not so simple. Sceptics might reply that this is no more than can be expected from a multinational enterprise like the oil industry, whose primary function is the creation of surplus value. Yet the Norwegian experience has indicated that a concern about the psychosocial consequences of offshore employment, as well as the socio-technical aspects of work-place relationships need not be incompatible with the profit motive. Indeed, the Norwegian sector of the North Sea reveals a variety of fruitful collaborations between public and private institutions, including government, the operators, trades unions and the research community (see Chapters 2 and 5).

Further comparisons between offshore oil developments reveal some interesting differences. Work scheduling is a particular example. In our study, conducted in an established production phase, 64 per cent of the women had partners who worked offshore on a regular work cycle of some kind; the remainder worked offshore intermittently on various forms of call out system. Of those women reporting regular absences, 65 per cent encountered cycles of two weeks on/off, 23 per cent one week on/off and 12 per cent four weeks on/off. In the Norwegian context however, also in the production phase, the vast majority of workers are employed in a regular work cycle of two weeks offshore, followed by three weeks at home (see Chapter 5). By contrast, in the exploration phase of the Newfoundland offshore, rig workers are usually employed for three weeks on/off. What do such figures imply? There is considerable variation in *families*' experience of offshore employment, though the evidence suggests that most people are prepared to put up with what they have got; indeed they may often be unaware that there are alternative systems. In the Scottish study, for example, some 73 per cent of the women when asked to choose a preferred work cycle opted for either one or two weeks on/off on a regular basis. By contrast, in the Newfoundland study (see Chapter 6), almost 70 per cent of male offshore workers had a preferred work cycle of three or four weeks on/off.

It would appear that differences in rotational cycle are related to local conditions and contingencies. There is little sense in which any of these (other than in Norway) take into account the psychosocial consequences of such work patterns, either for employees or their families. There appear to have been few attempts, for example, to study the long-term psychological or social consequences of twelve-hour work shifts carried out over regular periods of work offshore followed by rest periods onshore (Hellesøy, 1984, is an exception). The implementation of such patterns is clearly related to organisational, technical and economic concerns which take precedence over those of the employee, who is left with the responsibility for developing strategies for 'coping' with the consequences both in the work-place and at home.

There is also evidence to suggest that these work cycles shape other aspects of the lives of those involved, in particular the female spouses of offshore workers. Our study indicates a direct relationship between husbands' employment offshore and wives' participation in the labour market in the form of a clear variation in female labour force participation, according to whether or not the woman is married to an offshore worker. As we can see from Robert Moore's analysis of the Aberdeen Labour Market Area from the 1971 and 1981 Consensus (see Afterword) levels of paid employment among the women in our study (34 per cent) contrast sharply with that of women in a similar age group in the immediate Aberdeen locality (65 per cent). We might offer two competing explanations for this difference and hold them in tension. On the one hand it could be argued that women who are married to offshore oil workers are 'married to the job' to the extent that their chances of sustaining paid employment outside the home are seriously diminished. For these women the additional burden of responsibilities created by their husbands' job means that they are denied access to paid work for themselves. Alternatively, we could argue that for some women, the financial rewards of offshore employment mean that they need not consider a paid job of their own if they do not wish it. For these women, certain aspects of the 'corporate lifestyle' of the oil industry may have their attractions. We shall return to this later.

Offshore work obviously has numerous implications for various aspects of marriage, family and individual life. Before exploring these at greater length, however, we shall spell out the background and methodology to the Scottish study. In doing this it is useful to consider some of the underlying assumptions which were contained in the study design, and also to note those of other professionals who have an interest in the subject. In each case these assumptions serve to expose some taken-for-granted beliefs about the relationship between work and the family and also indicate ways in which the relationship may be problematised.

ORIGINS AND METHODOLOGY OF THE STUDY

It was in the mid-1970s that one of our colleagues, a consultant psychiatrist in Aberdeen, first became interested in the psychosocial consequences of offshore employment. This interest was based upon clinical observations of women who were married to or living with offshore oil workers and from whom their partners' intermittent absence and presence appeared to create special problems. These difficulties were seen in particular in the form of a triad of symptoms – anxiety, depression, and sexual problems. The triad came to be known as the 'intermittent husband syndrome' (Morrice and Taylor, 1978). It was acknowledged that such symptoms could occur among women whose partners were not employed in occupations involving absence; 'but when they occur in the circumstances described, they are a recognisable psychosocial entity, which deserves treatment and prevention' (p.13).

It is important to recognise, therefore, that the Scottish study reported here, inspired by these early clinical observations, was rooted in a *social problems* perspective. Initial descriptions of the 'intermittent husband syndrome' met with considerable local endorsement; general practitioners, health visitors, social workers and even oil company medical officers claimed to recognise the 'syndrome' as something they encountered in their own practice. After publication of the initial paper the 'syndrome' gained further currency and was often referred to in the press, television and radio. This endorsement and coverage gave further strength to Morrice and Taylor's argument for a research project to establish the incidence and prevalence of the 'syndrome' and make comparisons with other industries. Such a pattern has some of the hallmarks of the process known as 'deviancy amplification' (Cohen, 1972).

Appropriately qualified moral entrepreneurs draw attention to some problematic situation or behaviour, about which 'something must be done'; this leads to further debate, refinement of the problem category and the development of strategies for action. The process may gather a momentum of its own, allowing those involved to quickly lose sight of the possibility that the 'problem' in question may itself be problematic. And so it was with the 'intermittent husband syndrome'.

To postulate such a syndrome is clearly to make certain assumptions about the position of women within marriage and family life. It is to assume that the psychological and perhaps physical health of women may vary in relation to the nature and characteristics of their husbands' employment. As we have already suggested, this may be a valid assumption. It would not be difficult, however, for some slippage to occur in the argument: women's incorporation into their husbands' work produces certain consequences is one way of conceptualising the issue, but to go on to see the *women* as the problem category is obviously quite another. It is here that we run into difficulties between disciplinary perspectives. For whereas sociologists may be inclined to view a subject such as intermittent spouse absence as a feature of particular sorts of relationship between the spheres of employment and the family, it is more likely that those with an interest in psychological medicine and psychotherapy will incline to an individual perspective. Whereas this latter may certainly *take account* of external, structural factors, its point of departure will be to look for successful and unsuccessful adaptations within the framework imposed. In other words, while one perspective might wish to problematise 'offshore employment' as the object of enquiry, the other will look for variations in individual 'coping strategies'.

Such tensions were present as we went on to plan the research project as a collaborative interdisciplinary enterprise. They remained with us throughout the design phase and on into the data collection and subsequent analysis. Clearly, they were never fully resolved and variations in emphasis and interpretation may be detected in each of the three earlier papers which present findings from the study (Clark *et al.*, 1985; Morrice *et al.*, 1985; Taylor *et al.*, 1985). The present contribution is no exception to this process. Furthermore, it has been specifically influenced by the involvement of one of the authors in the Canadian study reported in Chapter 6.

Turning to more practical questions, we also encountered

considerable difficulties in identifying and gaining access to an adequate sample of women within the target population. These problems have been described in detail elsewhere (McCann, *et al.*, 1984) and were in large measure a product of the unwillingness of companies and operators within the industry to cooperate with our venture. We were likewise reluctant to engage in 'snowballing' techniques of sample generation, sensing that this might systematically exclude certain categories – particularly those women who were not a part of the offshore 'network'.

There was also the issue of whether or not comparisons with other industries should be included in the design. In the event this question was answered for us; resources did not permit us to identify comparable samples from fishing, long-distance lorry driving, or similar occupations. Instead we focused on comparisons *within* the oil industry, between women whose partners worked offshore and those whose partners did not. A telephone screening technique (McCann, *et al.*, 1984) was used to overcome our sampling problem and to make contact with each of these groups. A total of 431 postal questionnaires were sent out; 286 to the partners of men working offshore and 145 to a comparison group whose partners worked onshore. Completed questionnaires were returned by 200 of the offshore group and 103 of the onshore group, response rates of 70 per cent and 71 per cent respectively. The postal questionnaires contained a number of psychosocial and health measures (see Taylor *et al.*, 1985), along with a variety of questions which sought to explore the experience of family life in an oil household. Subsequently, twenty of the women in the offshore group and seventeen of their husbands took part in open-ended, tape recorded interviews in which they were invited to elaborate on some of the themes of the questionnaire. Drawing here on both the postal questionnaires and the interviews, we shall attempt to tease out some hitherto unexplored aspects of our data. It is clear from comments on the questionnaires as well as from the interviews, that many of our respondents had answered our questions in relation to a set of background assumptions about the oil industry and their place within it. Having examined some of the main findings from the survey we shall move on to consider accounts of family life in oil households, focussing on the special circumstances created by the continuous pattern of partings and reunions which is created by offshore employment. We shall then look at how this relates to the women's self image and to their perceptions of lifestyle. In considering any of these factors it is worth remembering that the concept of the 'intermittent husband syndrome' had been widely canvassed prior to the start of our study and that several of our respondents were clearly familiar with the various arguments.

DIFFERENTIAL REACTION: A STRUCTURAL APPROACH

While the study had its origins in a small series of clinical cases, it was never intended that it should be restricted to a study of prevalence of 'caseness' in the population. We were also interested in ascertaining whether some groups of women with intermittently absent husbands were more likely to be affected than others. For example, we wanted to know if local wives experienced fewer problems than those who had recently moved into the area and if those who went out to work found it easier than those who worked in the home. We also wanted to know about the effects of the husband's work pattern, and of the length of time the wife had been coping with intermittent husband absence.

We started with a number of assumptions about those in the 'offshore' group. First, that those less socially integrated, such as incomers and those without paid employment, would find intermittent husband absence more stressful than locals and those with a job outside the home. Secondly, that those with little previous experience of husband absence and those new to the oil industry would find intermittent husband absence more stressful than those with a long experience of the oil industry. Thirdly, that the husband's pattern of work would have some effect: irregular absences being more difficult than regular ones; and among the latter, longer absences being more difficult than those lasting only a week.

Before examining the extent to which these assumptions were confirmed it is necessary to elaborate on the data available for this part of the analysis.

The Zuckerman adjective checklist (Zuckerman and Lubin, 1965) was used as a measure of wives' anxiety levels when the husband was at home, and away. We found substantial differences between the 'home' and 'away' scores, and therefore used it as a measure of mood change associated with intermittent husband absence. Only twenty-nine respondents had identical 'home' and 'away' scores, 112 had a difference of between one and four and fifty-five had a difference of five or more. We took a difference of five or more to indicate significant mood change. This threshold identified 27.5 per cent in the sample as a whole, and our interest lay in the proportions identified in the various sub-groups.

Our measure of behavioural change was based on answers to questions about changes in sleeping and eating associated with intermittent husband absence. In the sample as a whole 59 per cent reported a change in eating and 61 per cent a change in sleeping patterns. Just over 40 per cent reported a change in both, and this constituted our definition of behavioural change. Again, our interest lay in identifying the proportion reporting behavioural change in each of the subgroups.

Our third measure of reaction to intermittent husband absence was based on answers to a number of questions about arguments and misunderstandings over the period when the husband was at home. A majority of wives (59 per cent) reported that arguments were usually provoked by one issue – mainly the limited time available for joint activities – and just over a third (37.6 per cent) reported two or more provoking issues. The presence of these multiple arguments constituted our definition of marital conflict, and our interest lay in identifying the proportion reporting marital conflict in each of the subgroups.

Considered together, these three measures provided an overall index of reactivity – of the wife's *direct* reaction to intermittent husband absence. Our health measures – symptoms, GP consultations and overall self-rating of health – represented an attempt to detect *indirect* effects. Our interest lay in determining whether those groups with the most pronounced reactions to intermittent husband absence were also characterised by poorer health.

Residence Status

Just over a half (54 per cent) of the wives were native to the Aberdeen area; just over a third (35 per cent) were immigrants from other parts of Britain and 11 per cent came from other countries, mainly the USA.

Comparing incomers with locals, there was no evidence to suggest that locals found intermittent husband absence easier. Indeed, for the only difference between the two groups which achieved statistical significance, it was the locals rather than the incomers who seemed to be most affected. On all other measures the two groups were remarkably similar. This rather surprising finding, in relation to our starting assumption, may be partly explicable in terms of group composition. Locals tended to be younger than incomers, were more likely to have pre-school children (61 per cent cf 39 per cent) and, for reasons not entirely clear, they were also more likely to be married to men who went offshore for longer periods (87 per cent cf 54 per cent). We found that each of these characteristics was associated with a more stressful experience of intermittent husband absence. The presumed protective effect of established social networks, which locals were more likely to enjoy than incomers, could therefore have been nullified by some of the other characteristics associated with residence status.

Employment Outside the Home

As we have seen, a third of the women were employed outside the home. Contrary to the expectation that those with outside work would find intermittent husband absence less stressful, we found that they experienced significantly higher levels of mood and behavioural change. Data from the intensive interviews confirmed that working wives found intermittent husband absence disruptive and suggested tht they experienced a good deal of role conflict. When their husbands were offshore the job provided them with a sense of purpose and a source of companionship; but when the husbands were at home, many wives admitted that they experienced difficulties in reconciling his demands with those of the job.

Despite these rather pronounced reactions to intermittent husband absence, there was no evidence of any effect on their health, either over the course of the last two weeks or over the last twelve months. Those employed outside the home also had a low GP consultation rate compared with their housebound sisters. This could, however, be explained by their greater difficulty in arranging and keeping appointments with their GPs.

Previous Experience of Husband Absence

Four out of ten wives (37 per cent) reported previous experience of regular husband absence, mainly as a result of service with the armed forces, the merchant navy or the trawler fleet.

On a review of all the evidence on mood and behavioural change it was clear that they found it easier to cope with the intermittent absences imposed by offshore oil work. We would have been surprised if it were otherwise, but we also found that their advantage stemmed from a number of factors not necessarily connected with previous experience. Comparing those with and without such experience, the former had been married longer (92 per cent at least five years cf 67 per cent of the latter), were less likely to have pre-school children (33 per cent cf 67 per cent) and their husbands were more likely to work offshore on a regular basis (41 per cent cf 28 per cent) and for one week at a time (62 per cent cf 37 per cent). All of these factors, plus the vital importance of the previous experience itself, explained the comparatively advantageous situation of the 'veterans' in our sample. However, it should be noted that their advantage did not 'spill over' into health: there was no evidence that they experienced less illness than those with no previous experience of husband absence.

Duration of Marriage

One-quarter of the sample had been married for less than five years. One hypothesis that the newly married would find intermittent husband absence more stressful was confirmed. Compared with their longer-married peers, a higher proportion reported mood and behavioural changes and in addition to this more pronounced reaction to intermittent husband absence, there was some evidence that those who had been married for a relatively short period also experienced poorer health. Over the two weeks prior to the interview, a significantly higher percentage reported three or more symptoms. They also reported a higher rate of consultation, though this difference could have occurred by chance.

These fairly substantial differences were not solely due to duration of marriage. Apart from being younger than those who had been married for a longer period, those more recently married were also more likely to have pre-school children (80 per cent cf 43 per cent) and to have husbands with no previous experience of work away from home (87 per cent cf 55 per cent).

Stage of Family Development

Of the 186 wives with children, eighty-two (41 per cent) had at least one pre-school child. The expectation that such wives would find intermittent husband absence more stressful was only partly confirmed. Compared with those having older children a higher proportion experienced mood changes associated with husband absence, they also experienced more behavioural change and marital conflict, but these results could have occurred by chance. There was no evidence to suggest that wives with young children experienced more illness.

Husbands' Work Pattern

While the majority of wives reported a regular pattern of husband presence and absence, thirty-six per cent never knew when their husbands would be called offshore. As expected, such wives experienced more stress. Compared with those whose experience was of regular absences, a higher proportion reported mood and behavioural changes and multiple symptoms over the two weeks prior to the interview. As a group, those experiencing irregular husband absences shared no other general characteristic, so we concluded that these results were a true reflection of the different work patterns.

Duration of Husbands' Absence

Of those wives reporting regular husband absences, 65 per cent reported a two week, twenty-three per cent a one week and twelve per cent a four-week cycle. Comparing those on a one-week cycle with those on a two-week cycle there was some evidence that the latter found intermittent husband absence more stressful. A higher proportion experienced mood and behavioural changes and the percentage reporting GP consultation was over twice as high. These differences may not have been entirely due to the duration of husbands' absence. Those involved in the one-week cycle were more likely to be 'veterans', both in a domestic and an occupational career sense. Compared with those involved in other work patterns they were more likely to have been married five years or more (90 per cent cf 62 per cent) and to have had previous experience of husband absence (62 per cent cf 37 per cent). Given these substantial differences in the composition of the two groups, we could not attribute all the observed reaction differences solely to duration of absence.

Risk Profiles

Having completed our examination of some of the factors associated with differential reaction to intermittent husband absence, we wanted to specify the reaction for some groups in more detail. Figure 4.1 presents profiles of those groups which were most adversely affected. For each group we have shown departures from the sample mean on the six measures defined above. Departures above the line (which represents the mean or expected value) indicate that the groups is 'better' than the sample as a whole, departures below the line indicate that it is 'worse' than the sample as a whole. The procedure is fairly crude but it does provide profiles which can be scanned to give an overall impression of each group's reaction to intermittent husband absence.

Of the five groups profiled in this way it is clear that those at the top of the diagram reacted more negatively to intermittent husband absence than those at the bottom. The 'newly marrieds', 'working wives' and those experiencing 'irregular husband absence' all showed fairly marked emotional and behavioural reactions. Those without previous experience of husband absence and 'young mothers' did not. As far as marital conflict is concerned, we can see that it was the 'working wives' who were most susceptible; and they and the 'newly marrieds' also experienced a disproportionate number of symptoms compared with other sample members. Comparing the reactions of the two groups most adversely affected – 'newly marrieds' and 'working wives' – we can see that the former experienced substantial behavioural change but only modest mood change; the latter experienced substantial mood change but rather less behavioural change. We were unable to explain these different patterns.

Cumulative Risk

Throughout the preceding analysis all groups have been treated as if they were mutually exclusive. This is clearly not so since there was a good deal of overlap between, for example, the newly married, those with pre-school children and those with no previous experience of husband absence. We have suggested ways in which this overlap or multiple group membership may have contributed towards the observed group differences, but we have not spelled out

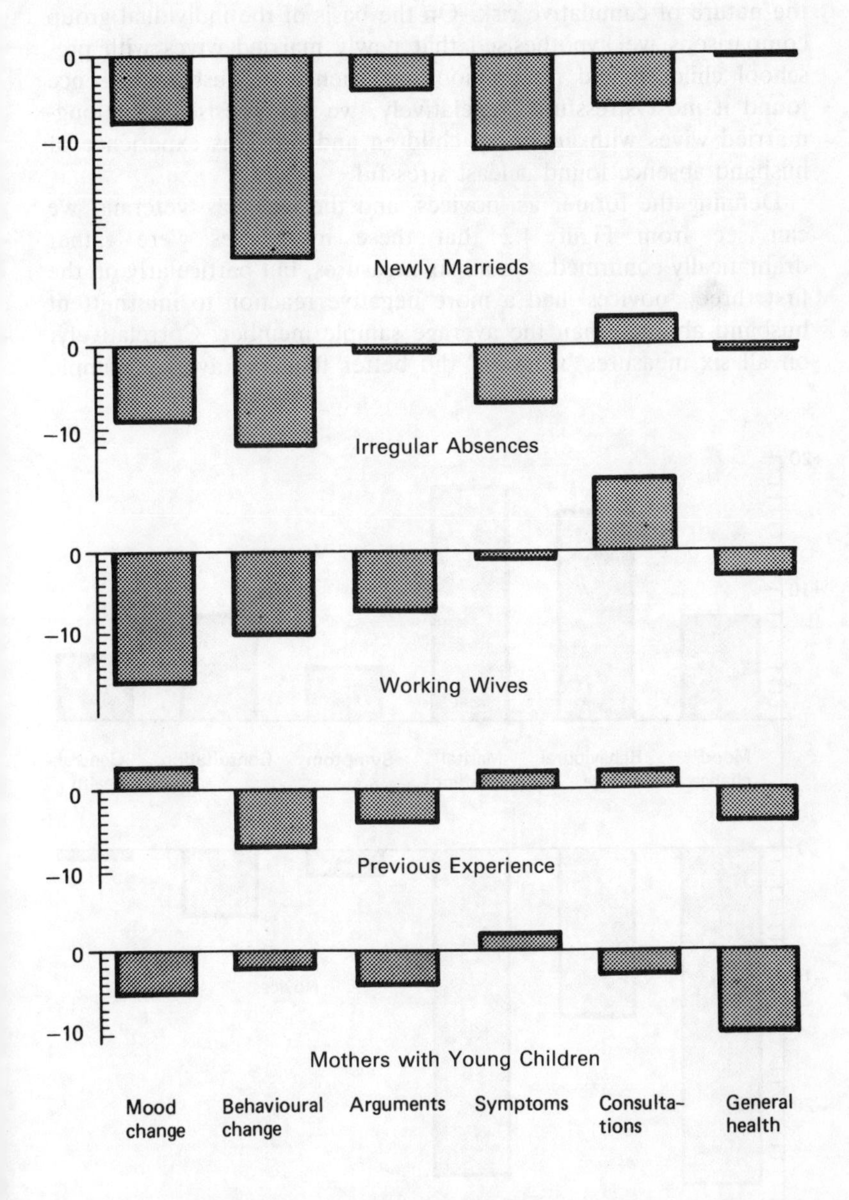

Figure 4.1 Differential reaction to intermittent husband absence: profiles for five 'risk' groups

the nature of cumulative risk. On the basis of the individual group comparisons we hypothesised that newly married wives with pre-school children and no previous experience of husband absence found it most stressful. Correlatively, we hypothesised that long-married wives with grown-up children and previous experience of husband absence found it least stressful.

Defining the former as 'novices' and the latter as 'veterans' we can see from Figure 4.2 that these hypotheses were rather dramatically confirmed. On all six measures, but particularly on the first three, 'novices' had a more negative reaction to intermittent husband absence than the average sample member. Correlatively, on all six measures 'veterans' did better than the average sample

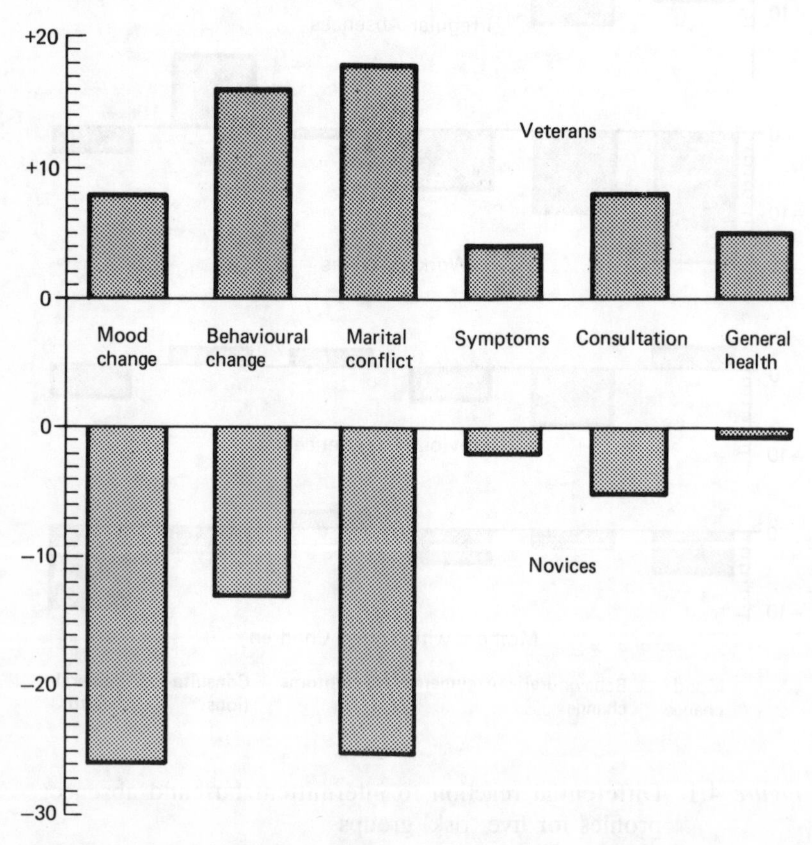

Figure 4.2 Cumulative risk: two extreme groups

member, in the sense that their reactions were less negative. The sharply contrasting profiles illustrate the difference between those who were and were not adjusted to intermittent husband absence.

They also raise a number of questions about the future. Veterans were undoubtedly better adjusted than novices when we interviewed them at a certain point in time. Would it be reasonable to expect that as novices gain experience, their reactions to intermittent husband absence will become more veteran-like? Without a prospective study it is impossible to know for sure, but we think that such a progression is unlikely. Those novices who find intermittent husband absence stressful have alternatives to continued stress. They can attempt to persuade their husbands to get another job not involving intermittent absences, and if they fail, they can leave their husbands. Given the nature of our sample it was impossible to know how often these alternatives were chosen, but many of the wives (and their husbands) admitted that they were only prepared to endure intermittent absences for a limited period. We had no means of knowing to what extent the divorce-prone stereotype of the industry is borne out, our sample consisting entirely of those currently married. Despite this lack of suitable data we view the veterans not as older novices, but as a self-selected group of survivors.

PARTINGS AND REUNIONS

We are aware that measures of 'reactivity' from the postal questionnaires can mask more subtle reactions, responses and adjustments which may take place within the oil family. These are particularly relevant in relation to the recurrent pattern of parting and reunion which is involved. Our data suggested that partings, and especially reunions, were sources of particular difficulties.

Most reunions were eagerly anticipated by both sides, the husband as well as the wife and children. They were invested with considerable significance, but frequently fell short of expectations. In the interviews the men reported extreme tiredness on their return home; some had worked a final shift immediately prior to leaving the platform, others found the helicopter flight stressful, even frightening. Consequently they were rarely in good physical or mental shape for the kind of homecoming which awaited them. One woman aged thirty-three, and at home with two young daughters, wrote the following remarks in her postal questionnaire in reply to a question about misunderstandings and arguments following her husband's return:

> it's more to do with mood changes I think. The children and I get excited when he is coming home and so we bombard him with the week's events. He's obviously unwinding too and gets flattened with everyone trying for his attention. He then gets short-tempered, especially if he's tired.

Accordingly, 50 per cent of the women reported that despite looking forward to their husband's return, some part of the stay was taken up with misunderstandings and niggling arguments. Almost a half of these women (43 per cent) said that the most common time for these arguments to occur was within a few days of his arrival back in the home.

In the interviews both husbands and wives reported difficulties in communication during this initial phase. There was a desire to 'catch up' on news and events as quickly as possible, cramming two weeks into an hour. At the same time many of the women felt that much of the day-to-day humdrum of child care, while reportable at the end of a 'normal' working day, appeared rather trivial and inconsequential at the end of a fortnight.

The communication of *feelings* was also difficult. Some women remarked in the postal questionnaire that neither they nor their partners could communicate at this level during the first few days. This was attributed to the men's tiredness, or occasionally to deeper reasons, as in the case of the following twenty-five year-old woman, at home with three children under age eight:

> Every time he arrives home, it's like having to start the relationship all over again. I find it hard to talk and I suppress my feelings. I feel like hugging him and telling him how much I love him, but I always see myself shying away and actually treating him like he has invaded my privacy.

Exactly a quarter of the total sample complained of difficulties which related to moods, fatigue or the emotions during the husbands' period at home. In some cases this amounted to resentment on the part of women who found themselves wishing their husbands were back offshore. One twenty-five year-old woman, whose husband worked four weeks on/off on a supply boat and who was at home with a small baby described this in her interview:

> I find myself not being able to be completely myself, completely relaxed. I don't feel able to do and say the things that I want to all the time. I find I'm sort of keeping myself in check . . . there are occasions during the four weeks he's home that I would rather he would go back to his boat.

There are clearly a variety of ways of understanding feelings of this sort in psychological or even psychodynamic terms. Morrice (1981) has argued that quarrels and recriminations among oil couples, and the tensions and tearfulness which promote them, may be understood as reactions to *loss*, akin to those seen in bereavement. Women whose husbands work away may therefore find partings and reunions problematic in so far as they continually interrupt the reverse emotional states. In this sense it is neither the *absence* nor the *presence* which is the cause of difficulty but rather the regular shift from one to the other.

Another way of conceptualising this is in terms of the oil worker's problems of transition between two *cultures* of home and work, each having radically different routines and value systems. One of the men described this in an interview in relation to the question of power and authority in the two spheres:

> the job I'm doing I have that sort of power to say 'Well? I don't care how you want to do it, you're going to do it my way' and then I come back here and you can't have that same authority, same dictatorship.

Another perspective on this was provided in an interview with a woman who described her husband's behaviour in the first few days after his return:

> he thinks he's still on the rig. He seems to be very heavy, you know. Where everything's quiet and peaceful in the house at night, he'll come in half past nine at night and I'll say 'be quiet, you're not on the rig', you know, and after that first two nights he's back to himself, he's fine again. . . I think the first couple of nights he's home, I feel he's noisy . . .

This contrast was particularly marked for those workers who had chosen to reduce travelling time by living near to the heliport. Within a few hours they moved from the extreme physical environment of the North Sea oil installation to the familiarities of the domestic world. For some the contrast was too dramatic and a 'buffer' was sought between the two, for example, in drinking bouts with work-mates before going home (for a more extended commentary on the role of pub culture, see Afterword). One man had relocated to Aberdeen from the North of England in order to avoid lengthy journeys between home and helicopter, but found later that he missed the extended train journey, which had provided an opportunity to relax, unwind and readjust to the onshore culture. Solheim (Chapter 5) elaborates this process in much more detail in relation to the Norwegian context and describes specific interconnections between platform culture and differing patterns of marital and family relationships.

A consideration of partings and reunions is therefore the starting point for examining more general questions about relationships in oil families. Since the departure and return are endemic to a way of life, they are given meaning within a broader context, one shaped by the prevailing value system of the oil industry. We did not of course set out to investigate such a value system, though our data does contain images of it, refracted through the prism of family relationships. In particular it is seen in the women's self-images, where it influences attitudes, reactions and adjustments to husband absence. It is also revealed in the perceived costs and benefits which attach to offshore employment. To understand such issues is to locate the 'intermittent husband syndrome' in a much more elaborated setting, which admits the possibility of individual 'coping' mechanisms, but at the same time points to other kinds of adjustment which might ease the transition between the cultures of home and work.

LIFESTYLES AND COPING STRATEGIES

While a small proportion of the women in our sample experienced severe mood and behavioural changes associated with their husband's absences, the majority 'coped'. We have already examined a number of structural correlates of coping; we now propose to explore the process from a perspective which takes account of the individual's place within the culture of the oil industry.

Positive Appraisal

It was obvious from the interviews that many wives coped by concentrating on the benefits, rather than the costs, accruing from their husbands' employment. Advantages were stated at various levels but one of the most frequently mentioned was that it permitted them to organise the time spent together in a way to suit themselves. For most this was viewed in terms of spending more time together than would be permitted if their husbands were employed elsewhere or, alternatively, that the time spent with one another could be used more fully. Most felt that the way their life was presently structured created a greater understanding and appreciation of one another. The following comments were fairly representative: 'We speak more freely to one another than we did when he worked at home. I'm probably closer to my husband than if he was home every night – this life makes you appreciate a good partner.'

Improvements in the marital relationship were also seen in terms of not taking one another for granted, certainly many felt that the space created by offshore working helped them to reflect more on their relationship and to place difficulties they experienced into perspective:

> I think you find that in your two weeks at home the quality of life is much better than what it was before when he came home the whole time, that is one of the big advantages. I find that quite often he's home for the whole fortnight and we never have an argument whereas before nothing major but just niggly little things, you know, we both tended to be a bit moody. If we had an argument it was nothing not to be speaking for a whole day and now, ehm, because you've only got fourteen days to give him you can't afford to waste a whole day not speaking, so that way its improved our life.

Some women particularly valued their husband's time offshore as providing an opportunity for independence and, to a lesser extent, solitude:

> I am able to do what I want, it has also taught me to be more independent. Thus gaining confidence and the ability to cope with guests socially and emotionally, try harder with people.

> You can be totally independent and follow your own interests and pursuits, also don't have to consider your partner if you want to visit, you just do what you want, therefore you feel more confident as a person.

Others also added to this benefit the improvement in their children's relationship with their father brought about by the longer, more concentrated periods of time father and children could spend together.

In addition to these advantages for family relationships, the relatively high financial reward was also seen as a positive aspect of the husband's job. Freedom from financial worries and a sense of general financial security was often thought to temper the difficulties experienced. Being married to an offshore oil worker had some of the difficulties of being a one-parent family (a number of women made this analogy) but without any of the financial worries. Some compared their present with their former situation:

> He was late on coming home and the kids were younger and going away to their beds early and he was away early in the morning. I just felt he didn't see his family, now, although he's away for a fortnight he's going to see an awful lot more of his family in his fortnight home.

The shadow of mass unemployment onshore also helped the women to see themselves in a privileged position: 'I left my family and friends but the way the job situation is I suppose we are lucky to have my husband working.'

Clearly, perception of the positive aspects of the job and the lifestyle form one of the means by which some women coped. This served to reduce levels of stress associated with the work offshore so that when crises occurred they were seen as minor irritations rather than the final straw.

Personal Resources

A sense of personal competence and ability was of key importance in relation to 'coping'. Frequently, women said such things as 'Nothing ever bothers me', 'I take it in my stride', 'I put my head down and keep going', 'I'm just that sort of person'. Thus, while some women could not articulate precisely how they coped, they attributed their success to personal strength and personality:

> I felt a bit strange to begin with, obviously when I found I was by

> myself and the kids, but I mean, I didna'. I knew that he was going and I'm the type of person to adapt to whatever's going on. Doesna' bother me at all . . . I knew he was away and that was that.

It was sometimes more of a necessity than a choice because of responsibilities for child care and the smooth management of the home. Children were often cited as a factor in continued coping:

> Life just goes on – I mean, there's no major difference because of the business, or because of the kids, some things are constant and you just have to get on with it whether he's here or not.

Self-esteem was also important. Perhaps the best illustration can be given by presenting a comparison of two women, Sue A. and Alison B. Both were in their late twenties at the time of interview, each with two school-age children. They had been living with or married to their husbands for roughly ten years. Neither was in paid employment and both had family in the area with whom they maintained regular contact. Their husbands had been working offshore for five years, they were both unskilled workers on a pattern of two weeks offshore, two weeks at home.

Sue A. had a poor view of herself. She felt useless and ugly. When her husband came home she felt she was boring because she had nothing to tell him about her life. She did little outside the house because she felt that there was very little she was capable of doing. She was very depressed and explained this almost totally in terms of her own inadequacy. Alison B. had a very different view of herself. She felt she could sort things out and was generally competent; through visiting friends and doing things she enjoyed, she capitalised on her husband's absence. She felt she was interesting, attractive and extremely capable. She could take anything in her stride. The substantial differences between these two women stemmed from different self-perceptions and they provide a good illustration of the extent to which personality is a fundamental determinant of coping.

Social Resources

Social support – family members or friends and neighbours – was invariably identified as a source of emotional and practical help. For example, one woman felt that without a close friend she would not have coped:

> Like I say, my parents are not feeling very well, you know, brother or sisters, whatever, we always tend to run for help to each other [close friend] sort of thing and get it off our chest and help each other cope with it. Like my dad died the year we came here. I don't think I could have gotten through without my friend.

Other friends in a similar position were identified as a source of assistance:

> Very lucky I am because I've got Diane who is also, she's got a husband offshore and, ehm, she becomes your confidante. I mean, anything that's worrying you I tell her and she tells me. This is the only way you've got a substitute [for husband]. If you were a woman on your own who maybe didn't have someone to talk to, then I think you really would be under a great deal or a lot more pressure.

In practical day-to-day circumstances, friends and neighbours played an important role:

> My immediate next-door neighbours are there if there's anything needs doing in the house he'll do it, you know, anything I can't tackle . . . If I want baby-sitters or anything one of the neighbours will help me usually.

Many of the women also insisted on the importance of developing an active social life. Around half were members of clubs, and the growth in organisations expressly for 'oil people' indicates the importance attached to informal social networks. The Petroleum Wives Club, which is made up of women whose husbands work in senior positions in the oil industry, is a good example. It organises a 'welcome wagon' to introduce newcomers to women in their area. Other informal 'welcome wagons' exist in most of the oil suburbs. There are also various expatriate groups. Americans living in Aberdeen have their own school, newspaper, club and sporting activities. So that when a husband is away, activity in these groups ensures that the women have continuing contact with others. This seemed to be especially important for foreign nationals: 'In this job situation whatever you want to call it you find that most expats stick together, they help each other out a heck of a lot, you almost get like a family.'

The Trinidadian expatriate community in Aberdeen replicated its own cultural traditions as much as possible, holding carnivals and parties and maintaining a sense of national identity:

> There are quite a lot of Trinidadians up here . . . At home it's carnival time now, right, February, we have a carnival at home so they are having a carnival party where you disguise and they play a lot of calypso and stuff.

This same women went on to say:

> You always seem to bounce upon someone you know. I don't know somehow they always get together, you know, because most of them are from a different country so everybody knows what it's like to leave home, to live somewhere else, kind of comfort each other, you know.

For those who were not a part of this international community, having family members in the vicinity was seen as valuable. While the importance of family often appear in terms of security rather than in terms of active participation, their role was undoubtedly beneficial, as one women remarked:

> And I've got my family I think that's got to help a lot as well . . . If I feel down at any time I just get in the car with my daughter, I usually go to my sister's or my mum's and just sit there, it's just the company I think I need.

This same women went around to her family for meals, her father mended anything that went wrong while her husband was away and her mother looked after the child while she went out to work. Another noted the importance of her family in preventing loneliness: 'Well, for the fortnight he's away I just, well, most of the family live in Aberdeen, so I just go and see them down town or they come up and see me.'

The presence of social networks, therefore, appeared to enhance coping by reducing the sense of isolation and providing both practical and emotional support.

Manipulating the Environment

Direct action to change external pressure appeared as a further means of coping. Paid employment served this purpose, with a few women actively seeking outside employment as a way to reduce the stress involved in husband absence:

> I deliberately went to get my job back when he went away. And it was only because I found it too much a whole day and a whole

evening, I found I was stuck in the house all the time. I had to get out.

Getting a job meant different things to different women. For some it provided a means by which they could actively alleviate problems associated with separation, while at the same time utilising their skills and providing the contact and social interaction, absent when their husband was offshore. For others it played an intrinsic role, shifting the emphasis from the husband's career and the demands made by his employment. For such women employment provided them with a sense of worth and independence:

> Most of the time I try to be cheerful, you know, I'm working and I find I'm always cheerful, always happy-go-lucky and that if I'm feeling low I just snap out of it because I know I've got to work, you know, your work keeps you happy, I think, meeting people and enjoying all that is fine.

Other women filled their time with frenetic activity. Decorating and refurbishing the house, cleaning, gardening, reading, all occupied the time while husbands were out of the way:

> I think if you didn't do your housework or sit and read or anything you would sit and mope and I suppose you would sit and think about different things. But looking after the house and the garden and different things, you just don't have time.

Another woman, when asked how she coped with her husband being away, said: 'Well, I keep terribly busy as you can see, I mean I've got four kids and, eh, before I started this business I was in Party Plan anyway, ehm, and I usually just keep very, very busy.' By keeping so busy, she found that she barely noticed that her husband had been away two weeks, so the strategy appeared to be effective.

Cultural Prescriptions Relevant to Successful Coping

The strategies we have outlined perhaps elucidate *how* coping is achieved, yet an equally intriguing question is *why* women cope? An explanation lies in an analysis of the cultural, social and economic milieu within which coping takes place. Finch (1983) has suggested an explanatory threefold model based on the primacy of the husband's career. First, it makes sound economic 'sense' for the husband to work, usually because he is at a higher skill level or has more training to offer and can command higher wages. Secondly, the current organisation of social life makes compliance easy and the developing of alternatives, for example the wife as the primary paid worker, extremely difficult. Thirdly, acceptance of the demands of a husband's career provides a concrete and comprehensive way of being a 'good wife'. Strong cultural prerequisites exist which suggest that women, particularly married women, should aspire to such a role. In developing this model Finch does not, of course, wish to portray women as passive individuals who merely acquiesce to the situation in which they find themselves, but to indicate the cultural prescriptions which shape a wife's involvement in her husband's work. This generalised model fits the situation of the 'oil wife' extremely well. The husband is the main breadwinner, his salary permits a very comfortable life and it is therefore 'sensible' for her to cope with any difficulties it might create.

> I don't know, I've always accepted it and I've always said to myself, well, I mean, he's got a good pay, you know, we've got a nice house and we've got a nice child, well, we've got our baby now. It's just something you accept you know, part of life now, just a job I suppose.

Moreover, many women considered it entirely reasonable for the wife to be supportive on the grounds that it might help advance the husband's career:

> I think the major thing is that my husband is wanting to work in the office and he's going to be gone all day, some nights and maybe weekends. Whereas we're used to this when he's home. He's got nothing to do at work for a whole week. It's going to take a lot of adjusting, oh, yeah, but he wants to do it and so I would not stand in his way.
>
> I never say I'm annoyed at his work because I think that would make him feel bad going offshore you know, so I kinda' keep it to myself, but then I just say, well, it's his job, you know.

In accepting the primacy of their husband's work and the necessity of his employment such women were able to minimise the stressful experience of their husbands' intermittent absences. The final context in which coping takes place is therefore a cultural one.

One of the ways in which we sought to summarise such fragmented comments, was by the use of three hypothetical 'vignettes' in the postal questionnaire. These suggested contrasting self-images of women married to offshore oil workers and pointed to broadly positive, negative and neutral responses to the oil work lifestyle. Roughly one-third of the women said they were 'most like' each of the following categories:

> Elizabeth can't get used to her husband being away. She feels that life is incomplete without him. She occasionally gets lonely and depressed and usually counts the days until his return. (32 per cent)
> Jill's life carries on pretty much the same whether her husband is at home or away. She has no strong feelings about his job and his absences offshore don't produce any great changes in her moods. (35 per cent)
> Janet found it rather difficult at first but now quite enjoys the times when her husband is away. She can follow her own interests and feels that her life is her own and that she has more confidence in herself. (30 per cent)

From our earlier remarks about novices and veterans, it will be clear that we think it unlikely that a linear pattern is involved whereby all Elizabeths become Janets. Clearly some do, but likewise some Elizabeths may quit or indeed continue to struggle through, in the absence of other alternatives.

CONCLUSIONS

In examining various aspects of the relationships between marriage and offshore employment we have given prominence to the recurrent partings and reunions because we feel that these processes encapsulate the central dilemma for many of the women involved. In our original clinical focus the regular partings and reunions were implicated in the anxiety and depression detected in a number of 'oil wives' seeking psychiatric treatment. From this initial focus on 'caseness' we have sought to show how the 'intermittent husband syndrome' can only be fully understood in a social and cultural context. A range of socio-demographic factors, including length of marriage, stage in the family development cycle and paid employment outside the home, have all been shown to influence the experience of intermittent husband absence. Organisational factors, such as the pattern and duration of husband absence, have also been shown to have an influence. In addition to these structural factors, we have seen that there are marked individual variations, relating to notions of self-image, self-esteem and personal coping style. Finally, we have sought to show how the experience of husband absence has been influenced by the culture of the oil industry and by generalised prescriptions relating to the role of the wife in contemporary society.

The latter point raises more general issues for our own and other studies described in this volume. There is a dilemma in engaging in certain sorts of research which might appear to buttress particular familial and economic structures. However much we might regard the work/family nexus in the oil industry as one which discriminates unfairly against women, we cannot expect it to disappear by some sociological sleight of hand. For some of the women in our study, intermittent husband absence clearly *was* a problem and was articulated as such. We have therefore responded to it at that level, showing patterns of variation in relation to demographic and personality factors. None of this is to deny that the 'problem' may be ultimately located in a much wider political and economic context.

Notes

1. The research reported in this paper was financed by a grant from the Oil Panel of the SSRC (now ESRC).
2. J.K.W. Morrice and Kathryn McCann also collaborated on the project.

5 Coming Home to Work: Men, Women and Marriage in the Norwegian Offshore Oil Industry

Jorun Solheim

INTRODUCTION

There seems today to be a growing public interest in understanding the relationship between *work* and *family*. This is not incidental, but related to a general process of social change in modern industrial society, whereby both institutions are undergoing a fundamental transformation. In this process, which is deeply bound up with changes in the roles of men and women in society, traditional lines of demarcation between work and family are tending to break down. This has led to an increasing awareness that these two spheres of life, so often considered as separate and segregated, are in fact intimately connected, and that the whole structure of work is directly related to the family system.

For historians and social scientists, this awareness of the connections between work and family can hardly be called new. There is a long tradition of theoretical interest in these issues, albeit few empirical studies of how the two domains really interact.[1] What is new, however, at least in the Norwegian context, is that in the worlds of public policy, social planning and business enterprise, there is a growing concern about how *specific* work organisations and work cultures affect family life, and also about how different family demands and expectations in turn affect the structure of work. Formerly the planning of most enterprises and work organisations were conducted without any special regard for compatibility with family life, which was left to adapt as best it could to a variety of work demands. Today we find the family and domestic organisation 'hitting back' in various ways on a work structure that is not tuned to its demands, and becoming recognised as a more active social force. The call for the six-hour working day is only one example of this.[2]

It is possible that this is a trend which is very much confined to the Norwegian, or Scandinavian, social setting. Still, I think this new recognition of the ways in which the family relates to the public sphere of work is part of a general development that, to some extent, can be said to apply to all Western industrial societies. Nevertheless, there are certain aspects of this process that seem particularly pronounced in Norwegian society and culture. Before analysing the special case of offshore work and family adaptation, I will outline some of the main changes that have recently been taking place in Norway in women's and men's relations to work and family life.

In the first place, while women are on the move *out* of the family, and into wider social spheres, men seem to be moving in the opposite direction, taking more interest in private life and personal development within the domestic sphere. This shift in the private/public dimension between the sexes is a tendency that has also been pointed out by some American writers (Kanter, 1977; Friedan, 1981), and it may be more or less pronounced in other countries. In Norway this trend may yet be reversed, and there are obvious differences in this respect according to locality and social class. But it is still an identifiable trend, and it is clearly shifting the balance in the relationship between the spheres of work and family.

The most significant development has been the steady increase during the last decade in the number of married women entering the labour market. During the period 1972–81 the employment rate of married women in Norway rose from 45 per cent to 62 per cent. Most of this increase, however, is in part-time work, and it may be argued that it has not therefore altered the relationship between women's work and family responsibilities in any really fundamental way. Still, the long-term *cultural* significance of this trend is likely to be profound. The occupation of full-time housewife is rapidly disappearing from the Norwegian scene, both as a personal preference and life-long vocation. This change is parallelled by the rapid rise of divorce, mostly initiated by women, and a growing cultural awareness in women of all age categories of the need to be economically self-sufficient.

For men, the movement *towards* the family is not as immediately obvious, since it does not result in married men quitting work, and becoming housefathers instead, although a few do. Still, there seems to be an increase in the ratio of men working part-time,[3] and there is also a growing complaint among married men in general that long working hours and job pressures mean that they have too little time to spend with their families. One significant aspect of this is the movement for the six-hour working day, which started out as a clearly defined women's issue, but which has more recently also gained some support from men, and is increasingly being looked upon as an opportunity for both sexes to combine work and family life in new ways.

The statistical evidence seems to bear out these interpretations of the broad trends taking place, and shows that the average time per day spent on wage work as opposed to other activities has altered significantly during the last decade. Employed men spend less time on work, and more time on household and family care, while women spend more time on work and considerably less on domestic duties. For both sexes the importance of *leisure* is clearly increasing (Table 5.1).

At the same time, the *total* amount of domestic work done in the family is declining. This, together with the increasing demand for leisure, points to another important cultural change, namely a shift in the whole concept of family life. While formerly the family existed as first and foremost a practical household enterprise, usually marked by different and complementary duties for men and women, the modern concept of family life is more detached from the household work partnership. The Norwegian family of today is increasingly seen as a special and exclusive sphere of personal

Table 5.1: Average time per day spent on different activities by men and women in Norway

	Employed men		*All women*	
	1971/72	*1980/81*	*1971/72*	*1980/81*
Wage work	6.5	5.7	1.9	2.4
Household work/family care	2.1	2.4	5.9	4.8
Leisure	5.0	5.7	5.0	5.9

Source: *The Time Budget Surveys 1971–72* and *1980–81*. Oslo: Central Bureau of Statistics, 1983.

intimacy and emotional support – a companionship of 'togetherness'.

Again, this idea of the family may vary considerably between different generations, classes and localities, but the emergence of a rather different family life style is still clearly identifiable. The modern Norwegian family puts a premium on time to *be together*, and the quality of the relationship between spouses tends to be measured in terms of personal involvement and closeness, rather than in terms of its effectiveness as a working partnership. Both these patterns of change, the shifts between the sexes in relation to the sexual division of labour, together with the emerging ideal of the 'companionate' marriage, have important implications for the relationship between work organisation and family life. Both patterns set quite new standards for the *compatibility* between the two spheres of work and family. The most significant aspect of both these processes is that they make possible a new familial role for the husband/father, which in turn may trigger off new balances and conflicts within the family itself.

In what follows, I will consider the impacts of the special context of *offshore work and commuting* on different types of Norwegian families, and also discuss how certain aspects of the family situation feed back into the offshore work community. The analysis is based on data collected by the Work Research Institutes in Oslo since 1981.[4] The families in the study are living in different local communities, and I will also discuss some of the main implications of locality on the way in which families deal with offshore work.

OFFSHORE WORK IN RELATION TO NORWEGIAN WORK CULTURE

The offshore industry is exceptional in that its work schedule may appear at first sight to be particularly compatible with the kind of changing life style and expectations I have briefly outlined above. The two-week work period followed by the three weeks at home, which is nowadays the normal offshore schedule in Norway, would seem to be rather well suited to a situation where being together as a family has become important value for men as well as for women. In fact, leisure time and time for the family is *the main reason* given by the married male offshore workers in our study for taking on such a job. In general, this factor seems more important than the wage (unlike the case of Newfoundland, see Chapter 6), and certainly much more important than any anticipated job satisfaction.

This, however, highlights a somewhat paradoxical situation, in that the value of the job is seen not to lie in the work itself, but rather in the *release* from it. This is aptly summed up in the common joke among offshore workers: 'The only problem with this job is the two weeks offshore'. This sort of attitude to the job tends to distinguish offshore work from more traditional occupations in Norway. In general, Norwegian work culture has been dominated by a relatively high regard for 'internal' values of the work itself, that is, job satisfaction, opportunities for learning and skill development, safety and general well-being. The Norwegian Work Environment Act of 1977 can be seen as a testimony to this, setting out, as it does, to secure that work should be both physically sound and safe, and socially and mentally rewarding. Even if the requirements of this Act are far from being met in much onshore industry, it still serves as a standard and an expression of important values in Norwegian work-life traditions.

By contrast, in offshore work the main values of the job are considered to lie totally *outside* the work context, in the form of external rewards, while the conditions of work itself tend to become secondary. This very marked 'instrumental' attitude to the work has in all probability not been an intended strategy of the oil workers themselves or of their unions. Rather, it should be seen as a defensive strategy adopted in a situation where the traditional norms of Norwegian work-life have been extremely difficult to uphold. The multinational oil companies on the Norwegian Shelf have had a long tradition of anti-unionism, and worker influence on job contracts and work environment has developed very slowly. (Liaaen, 1982; Karlsen, 1982; Qvale, 1985).

Furthermore, the oil platforms usually have a highly segregated work organisation, with a very centralised and bureaucratic structure of decision making. This means that there is little scope for influence over the work environment, as well as a fragmented and inefficient work routine. In general, these conditions affect both workers employed by the operating company and workers employed by the contractors, although the latter group have the least influence and usually poorer work conditions. (For discussion of the differences between these groups, see pp.64–5)

Because of these circumstances, much offshore work is experienced as dull and monotonous, with little scope for learning, variation or self-determination. On the other hand some work, particularly for contractor employees, is characterised by very severe time pressures and/or great physical strain, as well as exposure to health and safety risks. On the whole, work relations on the platforms are lacking much of the self-reliance and worker responsibility which is highly valued in traditional Norwegian work culture.[5]

It has therefore become an accepted tradition among offshore workers that the intrinsic values of the job have to be sacrificed in favour of external rewards in terms of money and leisure at home. This tends to create a spiral effect, where lack of identification with the work situation leads to an overload of expectations placed on home and family, which in turn may lead to further estrangement from the job (Holter, 1984). This brings us back to the question of *compatibility* between the two social settings, and how the social experience of the offshore period relates to the home and family situation. Before elaborating further on this, I want to outline those aspects of the offshore community as a social system which have particular significance for home life. (For a more detailed analysis of the Norwegian offshore community, see Chapter 2.)

THE SOCIAL CONTEXT OF OFFSHORE WORK

The platform community is a total institution, a twenty-four hour society, where social life is almost totally dominated by the work schedule. This does not mean that people are absorbed in work all the time, but that personal relations and free time on board are wholly subordinated to the organisation of work, and thus do not really appear as 'free'. Time, space and social action are highly structured and regulated, by the shift hours, the allocation of people to specific places and activities, and by people arriving and leaving the platform according to the fixed schedules. There is thus very little *choice* of social relationships, little variety in the content of these relations, and very few opportunities for leisure activity.

When, at the same time, the work relations are largely characterised by strain and subordination, and the work itself often appears as dull and monotonous, this lack of personal choice and self-determination becomes very pronounced. Taken together with the very real – and perceived – danger of working offshore, and the widespread feeling of being hopelessly trapped if things really go wrong, this situation adds up to an overall feeling of lacking *control* over one's personal situation. Thus we find that even if people like their work, and enjoy the company of their work mates, life on the platform is still felt to be both confining, and entailing a profound loss of freedom, and 'real' social and personal existence.

This prison-like quality of the platform is heightened by the lack of contact with society, and its fixed position in the sea. At the same time, and especially in smaller platform systems, the common fate of being cramped together on an isolated island may also create a certain feeling of belonging, and a special form of social integration. Nevertheless, this sociability is clearly secondary to the 'real' social life onshore. As one contractor employee in our study put it: 'Out there, we don't really live. We just exist.'

Because people are not really in charge of their situation, there is little scope for active participation, and passivity, personal withdrawal and disinterest become common symptoms. Social interaction on the platform is therefore often automatic, with little emotional investment, and activities tend to become routinised and lacking in personal involvement. At the same time, the whole platform community is usually characterised by a certain intimacy, albeit often highly ritualised, with a lot of joking and jesting. This easygoing mode of interaction may be seen as a necessary lubricator in a social situation which on the whole tends to have an atmosphere of estrangement and boredom. Even if there are important variations in this respect between different work groups and also between different platform systems (Solheim and Hanssen-Bauer, 1983; Holter, 1984), this *'boredom syndrome'* has up to now been a very marked feature of most Norwegian platform communities. It is often described by offshore workers themselves in terms of apathy, lack of meaning, dullness, 'to go brooding', etc. Such estranged, spectator-like attitudes very easily feed on themselves, and lack of participation breeds further passivity and monotony.

This social emptiness of platform life, rather than the drudgery of the work load itself, is perhaps the most basic reason why the offshore period is commonly described as strenuous and stressful. Apart from certain groups, mainly the catering and drilling workers, who *do* have physically hard jobs, it is, in the main, a popular myth that offshore work is particularly heavy. For the most offshore personnel, it seems rather to be the *lack* of meaningful activity, the long hours of waiting, and strained work relations that wears people out.

Thus we find an overall importance attached to the time spent at home, where the real, active life is supposed to be lived, and compensation sought for the strains and losses of the period offshore. The important question thus becomes: how does the social experience of the offshore situation affect the *capacity* of male offshore workers to participate in the life at home, and what are the implications for the family and community life?

ABSENCE AND TOGETHERNESS – A DOUBLE TRANSITION

Most offshore workers report that they need several days when they come home – often up to a week – before 'getting into the new rhythm'. Reference to this transition period is frequently made by offshore families, and it is often said to be a problem which increases over time, getting more marked the longer the offshore job lasts. It is my contention that this problem involves more than just the shift from one rhythm to another and the need to get over the work strain. In fact, it is the whole social and personal context of the offshore worker that has to be reorganised.

For the offshore commuter, the period at home usually appears to be the polar opposite of the platform life. The essence of the period onshore is 'free time', which is not regulated and prestructured, and where personal choice and involvement, and the creation of relationships and activities are dominant features. This means that the change to home life amounts in many ways to a total reorientation of the person from one state of being to another – from 'being ordered' to 'creating one's own order', from routine actions and personal withdrawal to active participation and personal involvement.

This cultural transition, however, may be very difficult to achieve. What is of particular relevance here, is that the mode of social interaction on the platform tends to be at odds with the requirements of the home and family situation. The latter calls for a quite different sort of *social competence*, in order to deal with close personal relations and unstructured situations in a creative manner.

This lack of compatibility between the offshore culture and the home and family setting may be more or less pronounced, depending on what kind of social roles the returning husband is expected to fill. This in turn will differ according to variations in types of family organisation, and the standards of social behaviour in the local community. Before commenting on these variations, I want to draw attention to the other major elements in the transition process: the reorganisation of the social life of the worker's wife and children.

One of the main findings of our study is that the period of the husband's absence and the period of joint family life entail very different lifestyles and modes of organisation for the wife and children at home. The way time is structured in the household, the tasks to be done, and the social activities and relations of the wife, are all likely to change each time the husband leaves and returns. Thus it is possible to say that there are two distinct forms of family and household organisation in the offshore family, one when the husband is away, and another when he comes home.

This means that the wife also continually experiences a fundamental transformation of her social reality, which has its own implications for joint family life. One of the most marked features of this transition is a change in the balance between *autonomy* and *dependence*. Most offshore wives find that while there is clearly an increase in responsibility for child care and household management when they are alone, there is also more autonomy in planning, and quite often fewer tasks to do in the household. The burden of responsibility, which is often deeply felt in the period of absence, is thus not so much related to the practical work tasks, but is more a question of being morally responsible, and having to bear the whole burden of organisational management in the family.

This is most pronounced for women with small children, where the relative autonomy in the household is counterbalanced by an increased dependence on others for social support, both practical and moral. The organising of a reliable network for help and assistance with regard to children is referred to as a major problem by these women, and constitutes a constant source of worry. This is related to the fact that the nuclear family in Norway is rapidly becoming very isolated socially, and there are few solidary women's networks or other institutions to which the wife may turn when alone. Close kinship or friendship ties are usually the only solution when such support is needed.

Dependence on an outside support network often disappears when the husband comes home, but is instead replaced by new forms of dependence within the family. A dominant feature of the husband's period onshore is the explicit or implicit demand on the wife's time, in the form of a strong expectation that she will devote herself to home and family. This is in part an expectation common to both partners, and results from the shared belief in the importance of 'being together'. Still, it very often runs counter to the woman's need to seek relief from the responsibility of household and family duties after her solitary existence as a caretaker.

The heavy demand on the women's time and attention during the husband's home period, which we have found to be very general, also means that there may be a significant change in her social life and orientation from one period to the other. In most cases, we find that her extra-familial activities tend to be most pronounced in the husband's absence, and that she has to cut them back during the period he is at home. This applies both to participation in organisations and community life, and to personal networks and friendships. In such situations the husband's period onshore may be partly experienced by the wife as a period of loss of social contact, and therefore as a source of complaint. However, we also find the reverse situation, where the period of absence is characterised by isolation and lack of social contact. This is most pronounced in rural communities, where fairly strict rules of conduct control and restrict the social space of a married woman who is alone. Housewives with small children, wherever they live, cannot easily leave the home in any case. In such cases, the wife may have a great need both for the husband's participation in the family when he comes home, and for social activities that take her beyond the boundaries of the home.

The offshore family is thus faced with not only one, but two processes of transition – for the husband and for the wife – between different social realities and modes of existence. This *double* transition and discontinuity is not commonly recognised, either by the spouses themselves, or in the community and society at large. This is partly due to a general cultural conception of the family as a 'natural', organic system, following from the unquestioned continuity of women's family labour. But this misrepresentation of the family as an unchanging reality is also very much sustained by the special circumstances of the commuting process.

From the point of view of the absent male offshore worker, the family exists as a fixed and immutable reality. It is *there*, frozen in his imagination as it was when he left, at least that is the hope. Just like his Newfoundland counterpart (see Chapter 6) he has very little notion of the family actually changing its way of life during his absence. In the local community, people also tend to view the offshore family as fixed and unchanging. The real discontinuity of the wife's situation is masked by the fact that the husband is so often present. This creates an illusion of a 'normal' and very self-sufficient household. 'What does she need help for – he is hardly ever away'. seems to be a standard reaction of local residents.

The wife may be very aware of the transformation in her own life situation from one period to the other, and the problems entailed in this 'invisible' transition. But at the same time we have found that she usually lacks concepts to express this process of discontinuity, and therefore tends not to communicate it. This, I will suggest, is an example of how women's own experience tends to be muted by the dominant cultural model, and fails to find expression in ordinary discourse.[6] This means that the *totality* of contradictions inherent in the double transition process tends to remain hidden, and very often results in unresolved conflicts and frustrations for both husband and wife. In order to develop this theme, I will turn now to some variations in types of family structure and in local standards of behaviour.

THE 'TRADITIONAL' FAMILY[7]

In more 'traditional' families, where the spheres of activity for husband and wife are separated and often clearly segregated, marriage may be considered to be first and foremost a complementary work partnership, and family life is often subsumed under the task of *household management*. In these families, the lack of compatibility between offshore life and home life for the husband need not be so very marked. The wife's transition from one period to the other will also usually be less pronounced. The caretaking of children and the creation of family relations will be seen as the woman's responsibility both when the husband is away and when he is home. His family role, on the other hand, will be acted out very much in terms of economic support and practical activities in the household.

In such a situation there will exist a certain continuity in social performance for both partners. The wife will go on performing her usual activities in the family, while the husband may get into a new, structured work routine at home, doing maintenance and building work in the house, or finding himself other types of work projects. There need not be much pressure on his active participation in the *internal* family relations, he can be 'left alone', and both partners may experience such a family situation as reasonable and proper.

Many of the offshore families we have encountered, especially in rural communities, follow such a pattern without too much conflict. For the wife, the main difficulties in this type of situation will be experienced in terms of *external* relations. The most common problem is loneliness and lack of social support during the period of the husband's absence. Most offshore wives in such a traditional family setting are full-time housewives, and many of them live in a state of marked social isolation during the husband's absence. In such situations, where the women are confined to the home without a supporting community network, we find that depression, anxiety and worry are common problems. On the other hand, we also find in this category some of the women who may be classified as 'copers' (see Chapter 4). They tend to be older, with long-established marriages, and a strong social network independent of their husband. They are often active in community life and organisations, and have a well defined social life of their own.

For the husband in this type of traditional family, the main object is to have suitable work projects going on at home, so that he can fulfil the standard social obligations expected of him as a household partner. The preoccupation that so many men working offshore have with their projects at home may be seen as indication that this is perceived as their main family role, where skills from the work context can be carried over into the home without too many problems. However, the scope for home projects is sometimes limited, especially in urban situations, and in any case they are very difficult to prolong indefinitely. Furthermore, it is a fairly common experience among offshore workers that such projects fail to materialise according to plan and that the valuable time at home tends to 'float away' or 'run out through your fingers'.

This seems to be related to the general problem of offshore-onshore transition described earlier. Many offshore workers openly state the difficulties of making decisions and structuring their time at home, after being conditioned for a long period to an external structure which makes all the decisions *for* them. In such circumstances idleness may become a serious problem. When 'free time' turns to idle time, the result can be very frustrating for both husband and wife. If the husband finds no real alternative family role, the results may well be that he withdraws from the family, or that he hangs around without purpose. In either case the expectations regarding the period at home tend to evaporate and be replaced by a sense of failure.

There is some evidence among the families in our study, that men in such a traditional family setting gradually tend to develop new skills and interests in the household, taking on more of the daily housework. Our research data so far indicate that most male offshore workers do more housework after having started working offshore than they did before. The most common domestic tasks the men engage in are kitchen work and cooking, and some of the cleaning duties. This gradual shift may be seen as partly a result of the difficulty of sustaining a traditional male work role in the household during the long free-time period. It also seems that men's more unbroken presence in the home makes it much less 'natural' to accept the cultural premise of women having full responsibility for the daily housework.

The time men spend with children usually increases with offshore work. Most of the families in our study regard the father's new opportunity to care for and be with the children as one of the most positive outcomes of offshore work. Even if the spouses retain very different duties and caretaking roles in respect of the children,[8] the mere presence of the father seems to alter his place in the family in a significant way. He becomes *accessible* in a way that is similar to the traditional position of the housewife. As the women in our study very often commented: 'Before he hardly saw the kids. Now they cling to his legs wherever he goes.'

The extent to which such a new family role for the husband/father evolves varies considerably. With the exception of child care, we have usually found that the most separate and complementary family roles persist in the rural communities, but this tendency is by no means uniform. Age of the couple, duration of marriage and personal history also play an important part, and make simple generalisations difficult.[9]

THE 'MODERN' FAMILY

In the more 'modern', companionate type of family, the shift to new family roles is more clearly established. Here the ideal family relationship is one of equality, sharing and togetherness, and the husband is expected to take his part in the internal relations of family life as an active and more or less equal *participant*.

This type of family situation is usually one in which the wife does not see her sole responsibility to be that of a housewife. She will usually have a job, either full-time or part-time, she may be enrolled in some form of further education, or she may be engaged in other social activities outside the home that occupy much of her time and interest. These women will therefore often experience the period of their husbands' absence as particularly burdensome, because it contradicts the notion of family equality and makes them the sole caretakers and household manager.

In such a situation, the double transition process for husband and wife entails other kinds of contradictions and problems. While the husband's expectations will be to get home *into* the family, the wife's needs may well be to get *out* of it, and be relieved of domestic duties and child care.[10] There will thus exist a strong imperative for the husband to be willing and able to take his turn as a caretaker when he returns, and for him to resume his place in the more intimate network of family relations.

This, however, is precisely what he cannot easily do, at least not immediately, because it presupposes a total shift from the work-structured and impersonal offshore behaviour. As one production operator commented: 'I often feel I get the family pushed down my ears, and I think: if only there was a work-load I could start on'. Yet among all the men in our study, the man was perhaps the most committed to the role of 'housefather', usually taking over most of the family responsibilities during his period onshore. In general, we found that most of the men who were prepared to share more equally in the family tasks, had problems with the transition from the offshore work culture to the personal involvement of family life. Many complained of a sense of estrangement, and of feeling 'artificial' and out of place.

In such cases, there exists a real possibility of family conflict and frustration, especially since both partners also share ideals about 'togetherness', which assume even greater importance than usual because of the need to compensate for the husband's absence. For the wife, this togetherness very often *presupposes* that the husband is able to take his turn in the family, so that she may have time to spare after doing her job or other extra-familial activities. For the husband, the expectations of a shared family life are based on the premise that the wife will be present and available, smoothing his way back into the intimacies of family life.

The male offshore worker's ideal of what it means to 'be at home' may thus be very different from the wife's need to 'have him home'. In fact, we find that few men really become engaged in family matters on an equal basis with their wives, and fewer still take over the whole domestic field as 'housefathers'. It seems that when the husband does involve himself in the domestic sphere and internal family relations, it is still on the implicit assumtion that the wife will continue to take primary responsibility for the home.

This conflict between the expectations of husband and wife often becomes even more pronounced in relation to the issue of women's work outside the home. It is commonly assumed that the wives of commuting men have difficulty in being wage workers because of their household responsibilities in the husband's *absence*. In the case of offshore commuting, we find the contrary to be more true. Most offshore wives we interviewed state that the main difficulties of having a job with a husband offshore were not related to the period of absence, but to the period when he was at home.[11] (This is also borne out by the Newfoundland data, see Chapter 6.)

Still, the majority of offshore wives who were working outside the home, either full-time or part-time, preferred to work than to be a housewife all the time. We found no instance of a woman quitting work because of her husband's offshore schedule, but quite a few who had started working after the offshore engagement had begun. At the same time the decision to take a job usually entailed some negotiation between spouses, and many of the women stated that they preferred part-time work in order to minimise the possible conflicts caused by their absence at work during the husband's period onshore.

The typical problems of the more 'modern' type of family in relation to offshore work can therefore be said to arise mainly from the following contradiction: while the modern ideal of marriage presupposes the *shared* responsibility between husband and wife in the personal relations of family life, the social conditions of offshore work tend to counteract this. In addition, there is the likelihood that wives' ideas about what constitutes sharing and equality in the family will also develop very differently from what is the current opinion in the nearly all-male social environment offshore.

In many ways then, the social context of offshore work continues to be based on the idea of a 'traditional' family system, where the male worker is able to return to *be* in a family organisation which is already there *for* him. When this traditional family is no longer a reality, or is in a process of rapid change, which seems to be the case in Norway today, social strain and conflict is very likely to appear.

We found such conflicts to be most pronounced in families which may be characterised as neither fully 'traditional' nor 'modern', but which were in a process of transition or uncertainty somewhere *between* these two cultural models. In both the most 'traditional' and in the most explicitly 'modern' families, there existed more well-defined rules about what family life and division of tasks *should* be, and these couples seemed to manage to develop a more or less shared conception about care, responsibilities and standards of behaviour. In the more 'in-between' families, there existed different and often contradictory rules and standards, which made it harder to fulfil expectations, and which resulted in greater confusion and frustration.

If such conflicts are not resolved, the result may well be some more fundamental disruption of family life. It is in fact a common belief throughout Norwegian society that offshore families are especially prone to divorce. In fact, there is no firm evidence that divorce is more frequent among offshore families than among other groups of the Norwegian population in similar age and occupation categories (Holter, 1982; Hellesøy, 1984). However, the divorce rate of offshore workers is difficult to estimate, since the offshore population is very unstable, and incipient divorce may well be one of the main reasons for people *leaving* offshore work.

What is important, however, is that the belief in a high frequency of divorce seems to be most widespread among offshore workers themselves. Most of the workers we have encountered were convinced that divorce was almost endemic among them. Rumours of divorce and family conflict is one of the most common topics of conversation on the platforms, and it is significant that the rumours usually concern *the deserting wife*. This seems to indicate men's profound anxiety and fear of losing the support of their families because of the conditions of offshore work. The prevalence of the divorce myth may thus be more a reflection of men's growing uneasiness and feeling of not catching up with the changing family culture, rather than a recognition of social reality.

FORMS OF SOCIAL INTEGRATION IN THE LOCAL COMMUNITY

The nature of the local environment has a very profound impact on the offshore family, both in terms of the kind of support network and social stimulation it provides for the wife, and in terms of the opportunities it provides for the husband's extra-familial activities. At the same time, the internal relations of the offshore family also set their own limits on members' social participation in the local community.

As a general trend, we have found offshore couples to be very 'familistic' in their orientation during the husband's period offshore, investing most of their time and commitment in the sphere of family and household. This is especially true of the men, but also applies to some extent to the women.

We have already commented on the push-pull effect on the wife due to the strong expectation that she will be present at home during the husband's period onshore, which tends to create a kind of commuting structure of its own. This back-and-forth movement of the wife in and out of the family sphere, will usually be most pronounced in an urban context, where there will be a greater variety of social arenas and relationships in which women can participate as *individuals* during the period of their husbands' absence. Many of the urban offshore wives lead a fairly active social life when they are alone, both in relation to their jobs and other social spheres. Mechanisms of social control are more flexible in the towns, and women tend to experience their social commitments as a matter of personal choice and preferences, rather than the imposition of specific cultural rules.

In rural communities, it is the family rather than the individual which is the core of the social network, and the wife will be excluded from much of this network when she is alone. She may be engaged in certain types of community organisations: church, choir, and voluntary social work, but there is usually very strong social control as to what are legitimate activities for a single married woman. To 'roam about' without purpose is much frowned upon, and the women have to be concerned all the time about not appearing idle and 'loose'. Conventions regarding visiting and social calls are also heavily restricted by the familial rules of conduct in the Norwegian countryside. A single woman cannot easily visit a home where both husband and wife are present, at least not in the evenings or at weekends, times that are marked off as 'family time'. With the exception of very close kin, many rural women state that they cannot normally visit friends and neighbours when their husbands are away, unless the other woman is alone too.

The 'girlfriend club' is an exception to this pattern, and may indeed be a consequence of it. This is an institution we do not find in the same form in urban areas. It is formally constituted, usually with meetings every fortnight in the home of one of the members, with quite elaborate rituals in respect of food and hospitality, and attendance is obligatory. Most of the rural offshore wives we have interviewed belong to one such club, and this is one of the few social obligations which is kept regardless of the husband's presence or absence.[12]

However, what we find in both the rural and urban areas is that in most cases the wife will have a need for active social contact outside the family when the husband comes home. This may be either because she wants to continue her social life and not lose her network of friends, or it may be because she has been cut off from social life altogether. This need very often conflicts with the situation of the husband, whose main interest is to *stay* with the family and not get very involved in community life. We have found this restricted interaction with the community to be widespread among married offshore workers, although it is by no means uniform. This trend towards family exclusivity may be attributed to the special conditions of offshore life described earlier. The lack of personal closeness and the exposed, public nature of social interaction on the platforms, tends to foster strong expectations of the privacy associated with family life. Many offshore workers, especially on large platforms, express a very clear opinion of being 'tired of people' after the two-week work period.

But this partial withdrawal from the community is also based on other grounds. Many offshore workers come to regard themselves as being more or less excluded from the community where they live. This is often described as 'getting out of step with society', or 'falling out of the system', and is perceived as one of the great *losses* resulting from offshore work. The main explanation people give for this social cut-off is the discontinuity in social relations and loss of vital information about what is going on. This applies in particular to participation in more formal organisations, where there is a constant progression of activities and flow of information, and where continuity of attendance is of special importance. Here, however, we find a significant and somewhat paradoxical difference between various *kinds* of organisational involvement. It seems that it may be easier to maintain membership in organisations requiring a low one. This may be because extensive engagement on the part of the offshore worker means that the organisation will also be more dependent on him, and more ready to wait for him to keep him informed. Of course this depends too on organisational type, and presupposes a family structure which will allow for such participation.

The alienation from the community network is probably also related to the difficulties of converting the highly structured social behaviour of the platform into the relatively unstructured and 'free' situation at home. We generally find that *when* offshore workers transcend the boundary between the family and community life, it is usually into some kind of established system where there is a known and fixed place 'waiting for them' which is more or less prescribed and therefore not necessary to negotiate personally. Besides formal organisations where they have a 'post', this will usually apply to close kinship and friendship ties, where belonging tends to be securely tied to the person himself, and not to continuous achievement.

Such security and predictability of a fixed place, however, may also apply to more impersonal places such as bars and restaurants, where the social setting is very standardised, and where there are no particular demands on a specific performance. This sort of social arena is usually restricted to urban situations,[13] and will often be perceived as an extension of the offshore community. This is particularly true of the more established 'oil towns' of Norway, such as Stavanger and Bergen, where many offshore workers are concentrated in one place.

A final cause of the offshore worker's estrangement and social withdrawal from the community is that his social identity as an oil worker carries a kind of *stigma* in relation to many dominant values of Norwegian culture. This is partly related to the relatively high wage level, which tends to offend local standards of equality and collective social responsibility. (A similar tendency exists in Newfoundland, see Chapter 6.) It is also very much related to the length of the period away from work, which is seen by many industrial workers onshore as a privilege not shared by other working people. Most offshore workers find it impossible to explain that they do not in fact have *more* leisure time than other shift workers, but rather that it is just differently distributed.

The stigma attached to oil workers is also related to various myths about the reckless and adventurous offshore life. These myths are a persistent survival from the first 'Texas period' in the North Sea in the 1970s, and seem to be quite unaffected by the changed offshore reality of today. The offshore worker is thus commonly portrayed – in the media and in common talk – as privileged and irresponsible, reckless and slightly immoral, with lots of loose money and idle time.

This stigmatised identity does not provide a very good basis for social interaction. Many offshore workers find it very difficult to explain to others the reality of offshore life and work, and to make their work conditions a topic for serious discussion. The workers employed by contracting companies have special problems in this respect. They have poor job security, irregular work schedules and usually bad working conditions (all of which affect both themselves and their families), and they find that their unions onshore pay little attention to their difficulties.

The feeling of being left out and forgotten often results in bitterness and withdrawal from 'society'. It contributes to the privatisation of the offshore family and the strain placed on the family system as the sole support structure. On the other hand it may also contribute to the formation of special and segregated 'oil' networks onshore. Both outcomes tend to sustain and reinforce the stigmatisation and cultural segregation of offshore oil workers.

CONCLUSION

A major object of this paper has been to outline the implications for the family system of the particular condition of offshore life and work. I have argued that the specific social context of platform life, and the social realities of the family and local community onshore, are very often out of step with each other. The offshore worker lives in *two cultures*, two worlds, which are to a large extent incompatible, and where social experience is difficult to translate from one context to the other.

But – and this is another main point of the argument – the offshore wife at home is also split between two forms of social existence, a situation which creates its own contraditions. In a certain sense the wife is also commuting back and forth between a single and a married state; a shift which entails different forms of behaviour, different social networks and activities, and different values.

This means that there are not two, but *three* distinct social realities in operation: his offshore life, her single life at home, and the joint life of 'togetherness'. These have to somehow fit each other if the offshore family is to survive. This may seem a trivial statement, but in fact it is not very much recognised. Most people, including offshore families, portray the situation as a split between *his work* on the platform and *the family* at home. This means that the contradictions and discontinuities in the wife's situation are not really brought into focus, but tend to be conflated under the common heading 'home and family'.

To map out this double set of contradictions, his and hers, and see how they interact, has been the main intention of this study. To make these contradictions visible and socially recognised may be the first and necessary step towards any real solutions of them.

There seems to be two main changes that would lessen the social strain inherent in the offshore work and family situation. One would be to develop the platforms as socially more stimulating and 'normal' places to live and work, which in turn would ease the transition of life at home and relieve some of the pressure on the family. There are certain indications today that such a development is taking place. The Norwegian oil companies, particularly Statoil, are gradually promoting changes in the work organisation offshore, and seem to be more concerned about safety, job satisfaction and the general social milieu on the platforms. The recruitment of women to offshore work also contributes to a 'normalisation' of the platforms as social systems.

The other kind of change concerns the social situation onshore, particularly the social support and integration of the offshore family into community networks, and the division of labour and responsibility within the family itself. These are undoubtedly matters of more long-term cultural change, which may be promoted by a more extensive knowledge of the internal logic of offshore families.

Notes

1. Work and family, production and reproduction, have long been general themes of interest in social history and anthropology. Within sociology and economics, these themes have mainly been developed within the framework of Marxist theory, for instance in the theoretical debate about domestic labour and capitalist economy. The way the areas of family and work interact as social systems has not, however,

been closely considered, and the empirical material is scanty (see Hareven, 1982).

2. Other examples are recent provisions in work legislation for extended leave of absence for *both* parents in regard to childbirth, children's illness, etc., and the growing pressure on both government and private firms to provide adequate nurseries and kindergartens for their employees.
3. The percentage of employed men in Norway who are working part-time (under twenty-nine hours per week) has risen from 5.6 in 1972 to 7.4 in 1981.
4. The main object of this study, which has been carried out by Hanne Heen, Øystein G. Holter and myself, has been to explore systematically the relationship between offshore work and family/household organisation. The data for this study is therefore of two types. In the first place, it consists of case histories of fifty offshore couples, living in three different localities: large town, small township and rural community. This case material includes only the families of *male* offshore workers. Secondly, data has also been collected in the course of participant observation on the main Norwegian production platforms in the North Sea (Ekofisk, Statfjord, Frigg), in order to study work organisation, social integration on the platforms and social adaptation to offshore life in general.
5. The Norwegian oil industry was, for the first ten years of its existence, dominated by mainly American traditions of industrial relations and work culture, which have been regarded as alien by most Norwegians. This situation, which has generated much social and political unrest, is now gradually changing. Both legislation and the stated philosophy of the Norwegian operating companies (particularly Statoil) have in the last years generated somewhat different models of work organisation and labour relations (see also Chapter 2).
6. The concepts of 'muted groups' and 'dominant cultural model' are borrowed from Ardener (1975).
7. I here use the concepts 'traditional' and 'modern' in a very general and ideal-type sense, and not as strictly defined analytical categories.
8. When such a division of labour regarding child care prevails, it usually means that the wife is responsible for the feeding and clothing and the more intimate tasks of personal attendance, while the husband is engaged in the more general role of supervisor and playmate.
9. In general, older couples (here, between fifty and sixty years) tend to have more gender-segregated family roles. We have also found that when the spouses themselves have had mothers who were housewives (particularly in the case of the husband), this tends to militate against more gender-equal family arrangements.
10. This need to be 'free' from family responsibilities, may also be a characteristic feature of the housewife's situation, as we have found in many of the interviews with women alone. They will, however, seldom openly make this request to their husbands.
11. This is in accordance with similar findings from sailors' families, based

on a comparative research project at the Work Research Institutes, Oslo (Borchgrevink and Melhuus, 1986).

12. This may seem to counteract the earlier statement about there being few solitary womens' networks supporting the wife in the husband's absence. The girlfriend club, however, very seldom functions as a support structure, at least not at the practical level. Indeed, it is often seen as a mechanism of social control, asserting the women's responsibilities as housewives and mothers.
13. The rural social landscape in Norway is singularly devoid of public places. There is no village structure, no 'pub culture' as in Britain, and few communal arenas for informal talk and interaction (see Afterword for further commentary on this difference).

6 Family Members' Experience of Offshore Oil Work in Newfoundland[1]

Jane Lewis, Mark Shrimpton and Keith Storey

'I don't think you every really adjust to it – you accept it' (twenty-eight year-old woman, husband five years offshore).

In a recent survey of the family members' experience of offshore employment in the Province of Newfoundland, Canada, most of those interviewed, like this woman, questioned whether adjustment to the work was possible. Rotational work patterns and the cycle of husband presence and absence were considered to be facts of life; in the words of another woman, 'You just have to get on with it'.

Following the work of the Aberdeen study team (see Chapter 4), the Newfoundland research began by seeking to document the problems identified by the families of offshore workers, although the survey included men as well as women and was not based on the hypothesis that female family members were likely to be suffering from the medical condition labelled 'intermittent husband syndrome' (Morrice and Taylor, 1978). However, even this more cautious approach proved too simple. After all, the 150 women and 101 men who completed questionnaires were all, by definition, 'coping,'[2] in that they had neither separated nor left the industry; hence the feeling of impatience expressed by many respondents, especially during follow-up personal interviews, when asked about adaptations and adjustments: it was felt that either you accepted offshore work and its problems or you did not.

Nevertheless, the research elicited considerable evidence to support the idea that well-defined areas of difficulty existed as a result of offshore work, especially in respect to partings and reunions; feelings of anxiety and loneliness; the sexual division of labour; and the pattern of communication between husband and wife. The respondents were aware of these and were prepared to discuss them, often at great length, despite the tendency of many to deny that there was anything extraordinary about offshore work patterns. The attitudes of both the men and women towards offshore work and its effects on family life may be characterised above all as ambivalent. Thus, any attempt to distinguish simply between those reacting positively and negatively to the lifestyle imposed by the work, or between those who are 'coping' well or ill, inevitably misses the complex conflicts in feelings and behaviour, and the complexity of the process by which these feelings are constructed.

Male rig and supply boat workers[3] and their wives showed considerable commitment to the industry. Only 18 per cent of the men and 19 per cent of the women hoped that they or their partners would not be working offshore in three years' time. Men and women value offshore work for its extrinsic rewards. It provides a well-paid job in a province where over 20 per cent of adult men are officially considered unemployed and where the real rate is considerably higher, especially in rural areas. Many respondents expressed an appreciation of the financial rewards and how they translated into new houses, cars and furniture. One couple interviewed, who lived in a new split-level bungalow, had a large coloured, framed print of the oil rig on which the husband worked on the living room wall. Having made the decision to work offshore, the majority of families feel that they must put up with the consequences. Many couples feel that it is not possible to 'adjust' to the continuous cycle of partings and reunions and its many ramifications for family and social life; as a way of life it will never be other than abnormal. Offshore work involves families in an ongoing, cost-benefit analysis, as to whether to continue working in the industry or to quit.

The tensions implicit in such a calculation are exacerbated in the case of the respondents to our survey, because their experiences are new in two important respects. First, they are participating in the early exploration phase of the industry in Newfoundland. Just over half of the male and female respondents reported less than five years experience of offshore oil work, although it is also true that over half the women have known no other type of work during their married lives. The sample, therefore, represents the first generation of families adjusting to this particular type of job and pattern of work. For while intermittent spouse absence is clearly a traditional feature of fishing and seafaring families, there are aspects of the culture of the oil industry which are relatively new to the Newfoundland setting. Secondly, the establishment of a repertoire of responses to spousal absence in the oil industry has been undertaken by a group of families who in general, have had little previous experience of either spousal intermittence or offshore oil employment in other settings. Their reactions and adjustments have therefore emerged *ab initio*, with little in the way of role models or examples of good practice.

Furthermore, the costs of accepting the family impacts of offshore work may not be equally shared between family members. In the sections that follow it is shown that while few male or female respondents have clear-cut reactions to offshore work, it is apparent that women are considerably more ambivalent about it than men. It is further argued that this can only be understood in the light of an analysis of the process of continual adjustment to spousal intermittence. For it seems that women must undertake a double adjustment to the pattern of work and the husband's needs that result from it.

THE SURVEY

The research strategy for the study was broadly modelled on that of the Scottish study and relied on both a postal questionnaire and personal interviews. As with the Scottish study, there were problems in locating the sample and, in the absence of a complete list of those offshore workers who were married or cohabiting, we used a combination of such names that were available, 'snowballing' and informal sources of information to generate a sample.

As the target population was identified, a three-stage survey was undertaken. First, telephone interviews were used to establish the usual rotational schedule, marital status, and the willingness of each individual to complete a postal questionnaire. Secondly, postal questionnaires were sent to all those indicating a willingness to respond in the telephone interviews. These contained a broad range of questions on oil work and family life. Finally, over thirty intensive open-ended, face-to-face interviews were carried out with individuals and couples who indicated a willingness to participate further in the study. The interviews were designed to probe specific feelings and behaviour during the course of a complete work cycle.

The population of offshore workers from which the sample was drawn was restricted to those rig and supply boat workers working on the Grand Banks and living in Newfoundland in July–August, 1985. At the time of the survey, the estimated offshore work force was 1756 with approximately 1018, or 58 per cent, of these workers resident in Newfoundland. Of these, approximately 502, or 49 per cent, were estimated to be married and 40.4 per cent of these workers indicated a willingness to receive a postal questionnaire.

Of the 203 questionnaires mailed to male and female spouses, 150 responses were received from females and 101 responses from males, which represent approximately 30 per cent of all female spouses and 20 per cent of all male spouses of Newfoundland resident offshore families.

REACTIONS TO OFFSHORE WORK

The primary measure of reaction to offshore work devised for the study was a self-classification scheme based on a series of three male and three female vignettes.[4] Each set of vignettes (see Figure 6.1) was designed to capture negative, neutral, and positive feelings about the offshore work pattern, with respondents being asked to choose the vignette that came closest to describing their feelings.

The vignettes proved a strong measure in discriminating between positive and negative feelings for both men and women. Only a small minority of men identified with the negative 'Nick' vignette; the vast majority identifying with the positive feelings expressed by 'Peter'. The women were much more evenly divided between the negative and positive characters of 'Nancy' and 'Pat'.

The men who have not got used to offshore work are distinguished chiefly by the fact that they are almost all rig workers, and by their negative feelings towards their work. Thirty-seven percent of all rig workers, as against 2 per cent of supply boat workers, classified themselves as Nicks.[5] The interviews revealed that the boats were perceived as less alienating, more familiar and more familial work places than the rigs, and this makes them

Nick cannot get used to the work pattern. He dislikes being away from his family. He occasionally gets lonely and depressed.

Mike's life seems to carry on pretty much the same whether he is at home or away. He has no strong feelings about being away from his family and his time offshore does not cause any great changes in his moods.

Peter likes his pattern of work. He accepts being away from his family but recognises the benefits of the time he can spend at home.

Nancy can't get used to his pattern of work. She feels life is incomplete when he is away from the family. She occasionally gets lonely and depressed.

Mary's life seems to carry on pretty much the same whether he is at home or away. She has no strong feelings about him being away from the family and his time does not produce any great changes in her moods.

Pat likes his pattern of work. She enjoys both the time he is away and when he is at home. She can follow her own interests and has more confidence in herself.

Figure 6.1 Male and female vignettes

Table 6.1 Male and female responses to vignettes

	M	%		*F*	%
Nick(−)	18	(18.2)	Nancy(−)	58	(39.2)
Mike(0)	5	(5.1)	Mary(0)	34	(23.0)
Peter(+)	76	(76.8)	Pat(+)	56	(37.8)
	99	(100.1)		148	(100.0)

considered more acceptable places to work by both men and their wives.

Safety was seen as a particularly important concern. As one wrote; 'Money doesn't mean much on a stormy night in February', which is likely a reference to the loss of the Ocean Ranger rig and its crew of eighty-four on the night of 15 February 1982. In response to a question asking them to give advice to an imaginary friend who was thinking of going to work offshore, one Nick wrote: 'Never go offshore because it isn't meant for humans.' Peters were generally more positive, often pointing out the financial rewards. When asked to provide advice on family life impacts to a hypothetical male friend joining the industry, the advice concentrated on exercise routines, safety precautions, and equipment. Seventy per cent of Peters wanted to be working offshore in three years' time.

Women's feelings about offshore work are determined both by their reactions to the nature of the work and by their husbands' feelings about it, for wives must also adjust to their husbands' moods and behaviour during their periods onshore. Not surprisingly, we found more Pats than Nancys to be the wives of supply boat workers, and more Nancys than Pats the wives of rig workers.

Like Peters, Pats tend to give evidence of having got much more used to offshore work and its shift patterns. Advising a hypothetical friend whose spouse was joining the industry, one commented: 'Keep a sense of humour and do fun things – the world doesn't stop because he is away.' It would seem that Pats are more willing to act independently, maintaining their own activities while the partner is offshore and making decisions by themselves if necessary. They are also much less prone to 'mood swings' between the times the male partner is on and offshore.

Nancys expressed difficulties in getting used to both the nature of the job itself and its implications for family life. One wrote that she would tell her hypothetical friend, 'that an increase in income does not compensate for the lonelines, fear, despair, and the absence of her husband'. Nancys worry about the danger of the job much more than Pats; they feel previous jobs held by the husband were less disruptive; and 95 per cent would like him to take an onshore job. With respect to family life, 33 per cent find that they do not see enough of their partners, whereas 13 per cent of Pats felt that they see *too much* of them. Nancys also find that much more disruption is caused to their social life and activities by the shift pattern, and while like Pats they agree that they have become more independent, there is the impression that this is largely because they have had to in order to survive. They find decision making more diffiult because the shift pattern stops them being together with their partners, and they were more diffident when asked if they agreed that in general the woman 'took over' when the man was offshore.

It is important to acknowledge the existence of inconsistencies within the vignette responses. Indeed, it would be incorrect to equate the idea of not being able to get used to the work with having wholly negative feelings about it, or of liking it with wholly positive feelings about it. Such an assumption is probably valid only for Nicks, whose feelings about the work seem wholly negative. The Peters, who tend to like the work, still bring up numerous negative points about it, especially when given the opportunity to do so in open-ended questions on the questionnaire. Pats also express a certain resentment about the additional responsibilities they have to shoulder when their husbands are offshore. As one wrote: 'You have to learn to be independent. You have to learn to be a single parent and sometimes do jobs your husband usually does such as mowing the lawn, banking.' This comes across clearly in the advice they offered their imaginary friends and their warnings as to the implications of having a 'part-time husband'. Pats have got used to offshore work, often after a period of considerable loneliness of the kind that Nancys continue to complain of. Another wrote: 'I would tell her that she will find the loneliness the worst . . . but after a while she will adjust as will her family but it will take time.' Their 'adjustment' does not necessarily mean that they feel wholly positive about it or necessarily value the greater independence they have achieved. Similarly, in the case of the Nancys, it might be expected that they would hope that their husbands would not be working offshore three years hence, but in fact they demonstrate considerable ambivalence, just as many Nancys and Pats recommended to their imaginary friends that the partner should accept the offshore job.

In part, such acquiescence reflects the benefits, particularly financial, that are generally perceived to accrue from offshore work, and in part, assumptions as to the proper roles of husband and wife in marriage. Despite their greater psychological dependence on their husbands, Nancys, like Pats, tend to agree that the choice of job rests primarily with their husbands. While a majority of women, 70 per cent of the total, and including almost all the Nancys, would like their husbands to take an equivalently-paid onshore job, of those expressing the opposite view, the most frequently given reason was that this was because the man liked his offshore job. Furthermore, a number of female respondents (11) said that the decision to give up offshore work would have to be the husband's. Similarly, when men and women considered the possibility of working in the offshore industry three years hence, those 38 per cent of women saying that it 'depended' seemed to think that it depended to a large degree on whether the man was contented. No man mentioned the issue of personal contentment with respect to himself or his spouse.

Many supply boat workers' wives felt that working at sea, 'is in his veins', as one put it. Another women, who identified herself as Pat wrote,

> I think the man's attitude about working offshore is very important. My husband works at sea and loves to be at sea. I do not think it would be right for me – his wife – to pressure him or make him feel guilty about being away from home when he enjoys his job so much.

Others felt strongly that the financial benefits of offshore work were important, especially given the high levels of unemployment in Newfoundland. These women also thought that the husband's willingness to take an offshore job should be seen as a willingness to fulfil his obligation to provide for his family. One Nancy advised her imaginary friend:

> you'll find it hard getting used to your husband being gone but just give it a while. I'd not advise her to tell her husband right away not to take it. It's hard on him being away but he's sensible thinking of the future.

A Pat was more forthright:

> I'd tell her she is lucky with jobs so hard to come by. She will get used to it in time and it will bring them both closer. I find that you haven't got time to argue and you get to go out more and enjoy yourselves when he do get in.

Most wives seem to see it as their duty to support their men on and offshore, as one woman put it in her questionnaire:

> an offshore worker leads two separate lives and in most cases one does not cross over to the other. One life is with his family and friends and another is with the 'guys' on the rig. Nobody really

> accepts this life style, just learns to live with it and the problems that seems to go with it . . . the offshore worker needs emotional support while on or off the rig and the person closest to him has to be the one to provide it.

Three women responding to the questionnaire summed up their feelings using the same phrase, 'If he's happy, you're happy.'

THE PROCESS OF ADJUSTMENT

Offshore work and rotational shift patterns require adjustment on the part of men with respect to the contrast between rig and home cultures; supply boats seem to present less of a contrast in this regard, even though the rotational pattern of thirty days on/off is longer.[6] Women must cope alone with parenting, domestic labour and in some cases paid labour, while the husband is offshore, with varying degrees of support from kin and community. They must also accommodate the moods and behaviour of an intermittent spouse. The couple must also work out roles, feelings, and obligations to each other and to children.

Men's Experiences

Even though the survey questions were designed to focus on the relationship between work and the family, male respondents wrote and spoke mainly about the work itself and particularly about conditions of safety and pay. All but two of the men responding to the question that asked them to advise a friend about to take a job offshore concentrated exclusively on the merits and demerits of the job, 'Put home life out of your mind, concentrate on work, take it one day at a time.' For most men the family is regarded as a relatively unproblematic 'haven' and valued as such. One man interviewed stressed the importance of a happy family life for rig workers, and this was also implicit in the dependence on their families many men seemed to exhibit as a result of offshore work. However, this was rarely openly acknowledged by the men, who tended to list the adjustments that were a necessary result of their intermittent absence, but who saw these as incumbent upon the spouse and family rather than themselves. Women and children were expected to adjust to the returning worker, rather than vice versa.

Responses to the open-ended questions in the male questionnaires were full of discussion of safety. This preoccupation was also reflected in the fact that 74 per cent of men think about danger 'sometimes' or 'often'. Men were also concerned about job security, employer/employee relations, benefits and, at the 'bottom end' of the work hierarchy (i.e. among those who are relatively unskilled and inexperienced), alienation from the job, which was expressed in the feeling of 'just being a number'. Two men referred to the rig as 'a prison' and to their need to 'let off steam' when they came home. Such responses were particularly prevalent in interviews with workers employed on American rigs; the culture of the rig, at its extreme during 'the American period', has been described elsewhere (Fuchs, Cake and Wright 1983, and Chapters 3 and 5) and similar reference was made in our survey to the way men felt that they had to keep their worries to themselves. One wrote: 'and don't be a person to be feeling down because this rubs off on fellow workers'. Personal problems had to be left at home because the rig was a 'different world . . . offshore is no place for someone who has problems at home'.

These themes were further elaborated in the interviews. One man referred to the need to 'live in the head [sic]' on the rig and his consequent need to relieve tensions at home. Another did not feel he could talk about family matters to other rig workers, who generally would not admit to any problems. Yet another said: 'I love to grumble when I get home first 'cause I don't have anyone on the rig to grumble at.' This man was content with the work, but another, expressing essentially the same sentiment, was not. He felt insecure and commented that you had to be 'a good con artist to get ahead' on the rig, the implication being that it was not possible to allow your true feelings to show.

There is also evidence that the rig might indeed be all-absorbing in terms of the workers' mental and physical energies. In an interview, one man complained that the approach to the work was dominated by time, rather than safety factors; the men were not treated like human beings and would sometimes have to work as they ate, or with injuries. There was conflict on the drill floor, with no respect accorded to lower grade workers and threats of demotion for those who were reluctant to comply. This man summed it up, 'It's not like going to work anymore, it's like going to war.' Another felt that it was not always possible to rely upon fellow workers' experience: 'If they need a guy to fill a position and they can't get someone to fill it, they'll take a green guy and put him in there right away and just let him figure it out.'

The sense of unremitting pressure was captured by another man who said,

> There's times when someone's standing right over your shoulder and keeps telling you to, 'do this, do this, do this'. The individual who's been doing the actual job, who's been doing that job for years, gets a little upset about it and things tend not to be as safe as they should be when that happens.

Perceptions did vary, however, and these appeared to be related to different drilling companies and management styles. Another man had not only found useful a preliminary session with his employer in which, 'They try hard to inform you about cultural shock, not seeing the family, and work', but also reported excellent work mates and high morale. There was also strong emphasis in some interviews on the buddy system, 'everybody got to look out for everybody else', though one man was careful when asked to say what he liked best about working offshore to distinguish between 'the people I work with, not the people I work for'.

In general, supply boat workers gave much more positive accounts of their work experience and talked about the benefits of a small work group and the cooperation and friendly relations with the captain ('the old man'). Factors such as this appear to make their thirty-day cycle more acceptable. By contrast, over a quarter of the rig workers who were interviewed would prefer to work fourteen rather than their current twenty-one days on/off in order to reduce the time spent in an environment which was seen as hostile and threatening. The third week offshore was seen by these men as particularly tiring, 'It's hard work and long shifts, in the third week the men's ability starts to peter out.' Some men described how this was exacerbated by a change from night and/or day shift in the middle of each offshore 'hitch' (work period offshore). There was speculation that both of these factors might increase the likelihood of accidents.

It is very clear, therefore, that coping with the job is uppermost in men's minds and this if anything increases their demands and dependence on the family. As one woman put it, 'When he's home he demands just as much attention and more than any child.' Men described the psychological excitement of reunion with their families in similar terms to those expressed by the women (see below, p.176) – a surge of energy for the one or two days before going ashore, the 'butterflies'. One man described this hyperactivity, 'The last two or three days I'm out here, *everybody* knows I'm coming home – I can guarantee you that.' Yet on reaching home this man described himself as often 'grouchy' and tired. The majority of men interviewed described their chief need on reaching home as sleep. This was particularly true for those working a full shift immediately before leaving the rig for home and those who live far from the heliport and for whom travel time is much longer. In the words of one of these men, '[when I get home] she's rarin' to go and I just wants to sit home'. Many men said forcefully that they did not want to confront household problems or necessarily to talk to their wives, although most were willing to put up with the children's clamour even when tired.

> When I comes home from the rig, I'm all wind up . . . twelve hours a day . . . you're steady winding. By the time you get out of there, I'm like a spring that's wind total. When I gets home I just want to [unwind] . . . sometimes I just even want to get drunk and forget about everything and wake up with a big hangover, then slowly break into the life at home. But if I comes home and here's a problem, right away I got to jump into another role – I'm the daddy now. I gotta take care of the son, or the wife needs comforting, or the sink broke down or something! . . . so you're playing two different roles, I find.

In a questionnaire, another man advised the hypothetical friend merely to, 'Get used to hearing, "I wish you had been here."' One man interviewed put it rather more forcefully, complaining that if his wife started to recount her difficulties during the past month, 'I don't need that. Do you know what I mean? After coming off – I just want to rest for a day or so. I don't need to be hit with something that's *her* job.'

Men working offshore have to do a job that most perceive as dangerous and many find difficulties in the transition between the 'cowboy culture' and home, dealing with it as best they can. Obviously, the shift from a world of men where 'you don't say "excuse me my dear"' to one of women and children is difficult. All look forward to coming home immensely, and expect the world of home and family to meet their needs and expectations; however, those of their wives may well be different. Most men seem able to establish a routine after a few days that involves fetching the mail (in the case of rural workers), working on some household project in the afternoon, doing things with their children after school, and visiting friends and family. Those who find it difficult to establish a routine, and who procrastinate with regard to home projects, also seem to be those who find the transition from one culture to another particularly problematic and frustrating. Most men interviewed were very family oriented when onshore and played a large role in caring for children. However, some men reported that they had had to give up community activities, such as dart leagues, because of their work schedules, which made them additionally dependent on their families for social interaction. One ex-fisherman who was interviewed said that he found that, 'offshore work puts you apart from your friends', and missed helping them with various projects, like house building. Only two mentioned nights out with the boys as a regular onshore activity, one of which involved socialising with other rig workers.

Most described feelings of irritability during the last few days of their leave. Many found parting from children particularly difficult; one said that before his children arrived, the rotational pattern had, if anything, served to relieve the boredom of marriage. Nor does it appear that parting from children gets easier: one man used a sports analogy to describe his feelings, 'Every time you go back to the rig it gets harder to get up for the game.' Another commented that as his son (four years' old) got older, he found it harder to leave him behind; on his last trip the boy had been 'screeching' in the doorway and he said to his wife, 'You know this is killing me, don't you?' Two men interviewed said that they did not sleep at all on the last night of their leave, staying up until they had to leave for the heliport at 4:00 a.m.; once back on the rig, they worked a twelve-hour shift, not having slept for forty-eight hours. Even in the case of workers who do sleep until leaving for the heliport, the first working day is long. For rig workers, the trip to the heliport and the helicopter ride are sources of additional tension. This may often spill over into the family, as one women said in an interview, 'The time that is the most stressful for me is on crew change days, because that involves the helicopter ride. I'm tense and on edge until I know the change has taken place safely.'

Supply boat workers seem to find going to the boat much easier. Three supply boat workers mentioned that the boats were 'more like a family' and that there was much less contrast between home and work. The tensions of partings and reunions were also eased because the supply boats spend time in port during the thirty-day work cycle, and wives and children (especially those living in or near St John's) may have the opportunity to see husbands and fathers at these times. Several women were also in the habit of taking their husbands back to the boat and spending some time with them onboard, drinking coffee and meeting work-mates and their wives. As a result, they felt more in touch with their spouses' place of work and the conditions. Two women pointed out, however, that despite visiting the boat in this way, they deliberately did not stay to watch it leave harbour.

Women's Experiences

Staying in the world familiar to both spouses, women find that they have a difficult path to tread in managing the tensions generated by the offshore work pattern. In the first place, women experience great anxiety as a result of the nature of offshore work. Of those saying that men should accept an equivalent onshore job, 17 per cent gave safety as the reason. All of the women reported that they thought about the dangers of offshore work, and almost a half said they often did so. Furthermore, 21 per cent said that they often talked about the dangers of the work in comparison with only 6 per cent of men. This seems to suggest that women have a greater need to express their fears regarding offshore work, a feeling which may not be shared by their partners.

The women interviewed reported feeling extreme tension, anxiety and excitement before reunions. One said that she was initially physically sick each time, while another used to break out in cold sweats as she drove to the heliport. Such extreme feelings may diminish over time, but the 'butterflies' remain. Many prepared for the husband's return with an orgy of cleaning and baking; one woman for example always made 'strawberry shortcake, cookies and a double-layer chocolate cake'. Often women's expectations of the husband's homecoming are doomed to disappointment. As one remarked:

> Yes, that really pisses me off. Here I am preparing for the great day when he gets home, psyching myself up for a good time and everything else, and then he comes home and all he can do is sleep for the first week. It's like a slap in the face 'cause the week before he gets home I say, Bill gets home in one week, three

days, two days – you know, count down. Then he gets home and all he does is sleep.

The man referred to in this quotation worked a twelve-hour shift immediately before leaving the rig and found that it took a week before he adjusted to a proper sleep pattern. Most took two to four days. This man also realised that many men did not want to find out what their wives had been doing while they had been away if it involved listening to problems, 'the attitude is, "well, I'm out there and I *work* for four weeks. When I go home I don't want to work. My work is *done*."' However, in spite of being able to recognise this as a problem he found himself unable to do anything about it.

Most of the women interviewed commented that they 'had learned' to respect the needs of spouses returning home for sleep, and to contain their desire to talk for at least a couple of days. For wives who have had to deal with a sick child or other domestic emergency, this may prove a strain; certainly this was the message expressed in three of the questionnaires. All wives recognised intuitively the importance of coming to terms with this problem of tension management if adjustment to the rotational pattern was to be made. One woman commented specifically on the time it had taken, and the importance of sorting out what was expected by her husband on the first few days onshore. In addition to sleep, a desire to get out in the open air and the woods was cited by some women as a typical part of their husbands' initial adjustment to the period at home. Another acquiesced to her husband's desire for sleep and sex, but still dreaded the first few days he was home.

When respondents were asked about arguments during the period the husband was onshore, 52 per cent of women and 40 per cent of men reported that some of the stay was taken up by 'niggling arguments'. Most of the women felt that these occurred either in the first or last few days of the period onshore, thus providing further evidence with respect to the difficulty of partings and reunions. One couple said that they felt tense for a few days before the man went offshore, with a sense of urgency and a need to get things accomplished. They would argue, 'over nothing'. However, the relatively small number reporting any difficulty perhaps reflects the degree of accommodation reached, primarily due to the effort made by wives. The causes of arguments given by respondents again suggest that it is indeed the wives who are primarily responsible for 'adjusting'. A majority of women reported that 'him adjusting to home' was the main cause of difficulty, whereas most men could not, or were unwilling to pinpoint anything in particular. As the wife of a supply boat worker wrote:

> I find it hard to cope with him at times in the month that he is home. It's a big change. He has just spent a full month doing his own thing with 14 men. Then he comes home for 30 days and he has to spend time with three kids and a wife. We have to change our lifestyle every month.

Most women perceived offshore work as hard and dangerous; even in an interview where the woman had forcefully expressed her own difficulties in dealing with an intermittent spouse, she none the less expressed the view that it was 'harder for him because he had to go away and leave the family'. Women's perceptions of men's problems were on the whole clearer than the men's perceptions of the problems of women. A particularly marked example of this occurred in an interview with a woman who was in full-time employment and who felt resentful that for her partner coming home was a change and a break. As she put it, 'I'm here day, after day, after day . . . it's new for him, he gets a break.'

Women find that they must work out how to deal with what for some is two separate lives (for further discussion of separate realities, see Chapter 5). To what extent to 'carry on' – seeing friends, going out, working – when the man is absent, or to put their lives 'on hold'? Some, from experience, are reluctant even under very adverse conditions to let their partners know about difficulties during the period while they are offshore. As one woman wrote:

> Things which happen at home with children or whatever if they have been really sick, the men know nothing of it until they come home, and then everything is over so he doesn't know what you really went through.

Certainly, more women than men found offshore work more disruptive than other jobs. In particular, husband absence poses problems with respect to social life, parenting and work routines, whether paid or unpaid.

Many women feel that they cannot attend social events without their partners. Most would never visit a bar alone or even entertain at home without their husband being there. Interestingly, few women indicated any frequent involvement in social activities with other women, for example, bingo or 'going out with the girls'. This is in contrast to other work on womens' culture in Newfoundland (see, for example, Davis, 1984). Thus, the greater the dependence on the husband for social life, the greater the problems experienced by the woman. Women who cannot drive or who have young children, may find themselves particularly tied to the home for practical as well as psychological reasons. One woman's children were unable to participate in scouts or other activities because she could not drive.

Most women seem to rely heavily on relatives for emotional and practical support during the period of spousal absence. From the interviews, it seems more are likely to visit relatives than friends and 68 per cent reported that they regularly got a relative, often a niece or mother, to stay overnight with them while their husbands were offshore. Few women suffered any stigma as a result of their husbands' absence and most seem to be able to count on a fair amount of kin and neighbour support; 86 per cent of women reported that they would turn to a relative for help if they were sick. If depressed, 48 per cent would turn to a relative and 49 per cent to a neighbour or friend. Professional help was considered to be of negligible importance in either situation. However, three women interviewed stressed the amount of jealousy among neighbours over the perceived wealth of oil workers. One of them also believed that her friends felt resentful about the fact that she saw less of them when her husband returned, and another respondent to the questionnaire felt quite simply that her friends just did not understand the problems raised by the periodic absences of her husband.

Only about a third of the women reported additional difficulty in dealing with children while their husbands were away. It was reported in the personal interviews that older children could more easily articulate their fear of the father's departure, which both parents tended to find upsetting. One woman expressed the fear that children would grow apart from their father. Another also felt that younger children feared their father when he returned, as they would a stranger. Mothers found difficulty in handling their own tension and children's anxieties at partings and mentioned that they experienced a few days' difficulty in dealing with children after the fathers left.

Much more problematic for most women was the issue of fathers missing important events in the children's lives like birthdays and particularly Christmas. One father had been home for only one of the last six Christmasses; another had missed the last four, and had therefore never spent a Christmas with his four year-old child. Many women found events such as confirmations and graduations difficult to attend alone. Children may, however, play a very positive role for women during the periods of husband absence. Three women stressed that they took great solace from their children during their husbands' periods offshore, and two childless women felt that children would have helped them to 'cope' better. One wrote: 'Because I have no children living with me, time is sometimes long after I get off work. So I must find some hobby to get interested in.'

Full-time housewives also had to cope with household maintenance, budgeting, etc., while their husbands were away and some obviously found that the responsibility weighed heavily. Others enjoyed the independence they had when their husbands were away and would probably have assumed the responsibility for the household regardless of the nature of their husband's work. For them problems arose only when husbands were unwilling to take a hand in babysitting and household chores when onshore and 'give the wives a break'. Two respondents to the questionnaire stated the problem forcefully, 'when my husband is onshore I want a break, but he wants it too,' and 'the man's view is that his time onshore is vacation time'. This woman felt that this was particularly unreasonable when offshore work patterns caused her great disruption in her own schedule. Most women felt that they should get all the housework done while their husbands were away, including tasks such as painting and decorating.

Only one third of the women worked for wages, compared to 48 per cent in the province as a whole. Women were glad to work while their husbands were away, but they experienced considerable difficulties in effecting the transition when husbands returned. The most successful in this respect was a substitute teacher, who could, if necessary, refuse work when her husband was onshore, but few others had this flexibility. One woman would have liked to have been able to take two days off when her husband came home but was unable to do so. Now that he came home on a Thursday, it was easier, because 'there was only one day to get through before the weekend'. Another woman who enjoyed her work and did not want to give it up, was experiencing considerable tension in her marriage because her husband resented coming home and doing housework and chauffering tasks. All the women interviewed who worked outside the home experienced some tension when their husbands were onshore. One woman who was attending university full-time commented that a lot went on in her life during a month and it was hard to catch up with her university work when her husband returned offshore. Undoubtedly, women who lead active lives while their husbands are away may find it more difficult than those who put their lives 'on hold' to reintegrate their husbands into their daily lives.

NEGOTIATING SPOUSAL INTERMITTENCE: THE COUPLE

Offshore work patterns also require adjustment on the part of the couple, with respect to the sexual division of labour, and power and decision making within the marriage. Indeed, couples require but do not always achieve, an enhanced awareness of each other's reactions to offshore work patterns and the needs arising from them.

In the majority of cases, men's dependence on the family increases as a result of offshore work, and some do play a more active part in domestic affairs; however, most are reluctant to assume any greater *responsibility* for household management. They do interact with children, but the key phrase in most of the interviews is their willingness to 'help out' with particular household tasks. Respondents to the questionnaire were asked whether they agreed or disagreed that they shared household tasks more equally as a result of offshore work, and whether they agreed or disagreed with the general statement that male offshore workers were more likely to help with housework. In both cases the men tended to agree. The women were somewhat less positive in response to both questions. In fact, only 29 per cent of men reported that they did housework often. There are undoubtedly still husbands who, in the words of one woman interviewed, 'don't know what a household chore is'.

At the other extreme, some women felt that their traditional role as carers and nurturers was somehow undermined by their husbands' absence. Two of those interviewed resented the husbands' praise for the food on the rig and had sought to reassert themselves by cooking all the things their men most liked on their return. One of these women also resented her husband interfering in the kitchen, which she regarded as her domain, while he was onshore, and described herself as 'a bit poisoned' when he came home and started to rearrange things there.

The vast majority of women felt that they had developed a greater sense of independence as a result of their husbands working offshore. This is undoubtedly fundamental to living with the offshore work pattern, and husbands of women who prove unable to do this to some degree are likely to leave the industry or separate, though we have no data on the prevalence of this. When asked whether they agreed or disagreed with the general statement 'when the man is an offshore worker it is the woman who takes over,' 59 per cent of women and 63 per cent of men agreed or strongly agreed. The interviews indicated that the couples in our sample, with one exception, shared the belief that the primary responsibility of the husband was to provide for his family, while that of the wife was to look after home and family. As Bell and Newby (1976) have pointed out, such a division of labour is usually sustained by a particular kind of 'deferential dialectic' between husband and wife, which is determined by the exercise of male power within the family. Thus the existence of both the need and the possibility of greater female independence during the partner's time offshore poses a threat to such relationships.

Men obviously think about the issue of what their wives do when they are away. Some men said that they did not see why their wives should not go out and many advised their imaginary friends to make sure their wives could cope alone, pay bills, and see friends. On the other hand, one woman respondent said that some men expected wives to stay at home and used the fact that they might telephone as an excuse. If a call did come through and the wife was not at home for whatever reason, she then felt guilty. Some men and women raised the possibility of sexual infidelity, also saying, inevitably, that it was not something that worried them personally.

Money management is a key item to be sorted out in the relationship, not least because all are agreed that money is perhaps the chief benefit of offshore work. Most women have to deal with money while the man is away, although some will leave even this until his return. A majority of those interviewed will turn the business of major bill paying back to the husband when he is onshore. Indeed, the majority of men and women agreed that there was no confusion as to who made the decisions as a result of offshore work. Nevertheless, while research has shown that there is a wide variety of methods of managing the family economy (Pahl, 1980, 1985; Land, 1983), very rarely do spouses of workers in other occupations have to reverse responsibilities on such a regular basis, and some tensions may result. One male respondent raised the issue of women spending the husband's hard earned money while he was away, while a female respondent wrote about men's drinking binges onshore which ate up savings.

Both men and women experience difficulty in adjusting to female independence. One man interviewed continued to keep control of particular jobs, for example, fixing the car, forbidding his wife to get repairs done while he was away and thus causing her considerable inconvenience. Some women also expressed the difficulty they faced in adjusting to their own greater sense of independence. The majority undoubtedly felt lonely and bereft, but one wrote to her imaginary friend:

> My advice would be to ask her husband to go to school if possible and train himself for some kind of job on land. They would be together as a family at all times and they wouldn't miss what they haven't got. The only reason for being offshore is the income. But as the years pass I feel that the time will come when I will be so independent that I won't ever want my husband to come home. Right now every time he comes home it upsets everything that I have organized and planned for that month.

Another began with forthright advice to tell the man going offshore that 'his time off is not party time' and to get a babysitter for a few hours twice a week while he is away to give herself a break, 'most of all – but not,' she added, 'separate from your husband'.

Thus, it is possible to see how the balance of relationships within a marriage must be renegotiated when one of the partners works offshore. However, communication betwen husband and wife is often problematic. One man explained the difficulty he experienced in this way:

> You can't really explain it [what it's like to work offshore] [they ask] what's the seas like? The rig was riding out 80–90 knot seas, 'What's that like?' They can't picture it. I told some of them, 'The Royal Trust building is 80 foot high – go downtown outside the Royal Trust building, and that's an 80 foot sea.' But they just can't picture it – they can't imagine something like that, right?

Some women would like to know more about their husband's work and two expressed resentment that they only learned more about it when their husbands were talking to others. But it is possible that the difficulties in articulating the nature of life on the rig are such that men give up the effort to explain. A significant minority of women do *not* want to know any more about work on the rig, largely for fear of increasing their own anxieties. Many do not talk about work to their husbands, for as one put it, 'when he comes off the rig, I figure he just wants to leave it behind him'. It appears that most men do indeed wish to do just that. Another common experience seems to be that talking about the anxiety does not make it go away: 'At first we'd talk about it, but it doesn't really change things or make any difference. It's better just to tell each other that we love and think about each other and not to let anything interfere.' This may nevertheless impose greater strains on women, who seem to want to talk more about safety than men.

As has been seen, men often do not want to face domestic problems and responsibilities when they return home and it can be difficult for couples to work out their respective needs. One woman wrote:

> It is difficult for [us] to work out the problems that we have like any couple. When he is in we don't want to go into problems. But then if we start, it seems it never can be solved. Dialogue is cut because he goes away. Everything has to be started all over again, everytime he comes back.

Women who wrote to their husbands were conscious of the problem that if they indulged their own feelings after parting and wrote a gloomy letter, it might well not arrive until the middle or late in the husband's hitch offshore and might then cause him to worry, possibly adversely affecting his safety.

Coming to terms with expressing feelings and resolving difficulties as a couple is difficult and this makes communication during the period while the partner is offshore especially important. A common pattern would seem to be that women write more letters, while men are more likely to telephone. In the majority of cases this would seem to be for logistical reasons and because the men fear that a call from home means bad news. Nearly all the women expressed some anxiety about safety and very few were happy with the communications systems provided by companies. Most felt that there was an inadequate mail service. Forty-nine per cent of women and 37 per cent of men wrote to their partners, but letters sometimes met the husband coming onshore, and most felt that they could not rely on mail delivery. One woman advised her imaginary friend that she should write a few lines each night 'like a conversation', but for many the poor mail service would render such an effort fruitless.

Telephone systems were also reported to be grossly inadequate. The rigs have both radio and satellite links with St John's and the provincial telephone network. The radio systems, however, provide a poor quality signal and no privacy (anyone with a radio receiver can listen in), while the satellite link is extremely expensive and some drilling companies severely restrict crew access to it. One woman described how, 'My husband has tried several times to reach me ship-to-shore and we couldn't understand what the other was saying . . . he called me on my birthday by Marisat (satellite hookup). We talked for five minutes exactly and it cost us $60. An expensive gift for five minutes of happiness on your birthday.'

Many also reported great difficulty getting reliable information on hitch changes. One woman told of driving to the heliport in St John's for 10:00 a.m. (when her husband was due) and waiting until 5:30 p.m. before being told that he would not arrive until 11:00 a.m. the next day. She drove home and came back the next day, and waited until 7:00 p.m. before he eventually arrived. As one man interviewed put it, it would be 'a common courtesy' to provide accurate information to families about shift changes. For women living in rural Newfoundland in particular, meeting a husband and father is a special event involving driving a considerable distance. The children's disappointment if the father does not arrive is great. One woman expressed her anger at the lack of information available in these terms:

> To me the people working there [the heliport] are a bunch of pigs and don't really know how much it means to the wives and children to know what's going on. I think there should be a phone on the rigs. A call from your husband once a week would mean an awful lot. I don't think that would be too much to ask.

CONCLUSION

There are, therefore, substantial costs to be shouldered by the families of offshore workers and these fall disproportionately on wives. This in large part explains the greater ambivalence of their reactions to offshore work. Women's acquiescence stems from both their assessment of the financial benefits of offshore work, which is shared by their husbands, and from their belief that they should not dictate a change in their husband's mode of employment, but rather provide them with practical and emotional support.

In undertaking their 'double adjustment' to the work pattern and to their husbands' needs, women appear to behave in one of three ways. Some women seem simply to 'carry-on', trying wherever possible to prevent intermittent spousal absence from interferring with their routines and concerns, perhaps with the support of a close friend or relative. A second option is to try to do everything during the period the spouse is away, including all aspects of 'his' role which may require attention; many of these women then go 'into reverse' on their partners' return – passing back to him the various duties and responsibilities which in their view are rightly and properly his. The third option is to place life 'on hold', to minimise all activities and concerns during the husband's period offshore in order to activate them fully when the spouse is at home. Although our study can say nothing about them there are, of course, two further alternatives for women who are unwilling to tolerate the way of life at all: they might get their partner to quit the job, or quit their partner. However, both are highly problematic in a province which presents few employment options and with generally traditional attitudes towards marriage and the family.

These patterns, however, are by no means so straightforward as they might appear at first glance. The woman's internal conflicts cannot be separated from those of her spouse, or from those which may be present within the family as a whole. She may, therefore, be pulled in several directions at once, by forces which vary in strength and persistence. This is not to deny that other members of the family are placed under similar strains, but the evidence suggests that it is the female spouse/mother who is most active in conflict management of various kinds. Of course, we know that this is a feature of wives in many family settings within our culture. We argue here that it is particularly marked and that in consequence, the notion suggested in the literature, that *wives*, should develop 'coping' mechanisms or become more independent is not appropriate. Reactions and adaptations should be seen as *family* issues and not reduced to the level of individual, i.e. female inadequacy, still less pathology.

Turning to the men, it is possible to identify certain features of their employment experience which highlight the special case of the oil family. Our evidence, as well as the Scottish and Norwegian literature, suggest that the oil industry is different in its effects from other, traditional occupations like seafaring or fishing, which involve intermittent spouse absence. Most particularly our comparisons of rig and supply boat workers bear this out. Employment on an offshore drilling rig appears to create certain unique experiences, producing individual reactions and adjustments. The men who replied to the questionnaires and who took part in the interviews were much preoccupied with the pressures of their work; they wrote and spoke little of their family lives. When they did, they referred to their own problems of adjustment, tiredness and irritability on their return. It should, of course, be recognised that for these men the culture of the work place, combined perhaps with the prevailing norms of the industry as a whole, are likely to enhance their self-image as 'breadwinners' doing an important and dangerous job in difficult conditions; those who engage in such work make personal sacrifices and should therefore be accorded certain family indulgences as a result.

To argue for such a distinction between men and women's responses to offshore oil employment is by no sense to descend into stereotype. Our study found little evidence of beleaguered 'oil wives' failing to 'cope' with the 'intermittent husband syndrome' while their partners enjoyed all the rewards of a well-paid job with considerable periods of leisure time. In this sense our findings reflect those of Clark *et al.*'s (1985) Scottish study in which couples portrayed themselves as in the main jointly and pragmatically engaged in a relationship wherein offshore employment was seen as a highly instrumental means to the achievement of certain material and financial ends. The men and women who took part in our study for the most part therefore placed their individual and family difficulties within a context which made considerable sense to them, as part of a package containing both costs and benefits which were readily acknowledged, especially by the women. In these areas, where the overlaps between the worlds of work and family are so clearly evident, there is room for changes in industry policies and practices which could lead to an improved set of circumstances both for those who are 'visibly', formally and contractually employed, as well as those women in particular who are the 'invisible workers' in Newfoundland's offshore oil industry.

While much of the literature on spousal intermittence suggests that women should be encouraged to become more independent (Boss, McCubbin and Lester, 1979; Patterson and McCubbin, 1984;

Voydanoff, 1980), it is clear from this study that this is an overly simplistic and inadequate response. The first focus of attention must be the effort to humanise the work culture, especially in respect to conditions on the rigs and platforms. The second must address the whole issue of power within the family. Despite the possibilities for men to share the domestic work of caring and household labour more equally, albeit on an intermittent basis, in practice it seems that men do not assume any greater responsibility for the domestic sphere, but rather add more needs of their own to the burden of caring that falls on women. As Hilary Graham (1984, p.185) has remarked, women 'tend to develop complex coping strategies [to] sustain the fragile equilibrium of their everyday life'. The way in which the meaning and range of women's choices is circumscribed by the organisation of family life is heightened in the case of offshore oil wives.

Notes

1. The research for this study was funded by the Environmental Studies Revolving Funds (ESRF), set up by the Canadian Oil and Gas Act and administered by the Federal Government of Canada. The research contract was awarded to Community Resource Services (1984) Ltd, whose principals are Mark Shrimpton, Keith Storey and Jane Lewis. David Clark, Doug House, Marilyn Porter and the Ocean Ranger Foundation acted as consultants to the project. For further details of the research see Community Resource Services (1984) Ltd (1986).
2. The definition of 'coping' used by the medical researchers working on Mobil Exploration Norway Inc.'s research project on 'Working Environment Health and Safety on the Statfjord Field' is, 'the ways people seek to manage or master stressful experiences so that a state of psychological equilibrium is established and generally maintained' (Hellesøy, 1981, p.13). However, Hilary Graham (1982, 1984) has identified 'coping with crisis' as an activity which underlies much of women's work in the family. She analyses the way in which coping is essentially self-help and how the idea of 'unobstrusive competence' serves as a yardstick by which women measure their own performance. Thus in Graham's analysis, coping can be health-threatening as well as health-sustaining.
3. Only thirty-seven women were employed offshore in late 1985 when this survey was completed and of these it is estimated that ten were married. None was picked up in our sample. On the impact to family life of women working offshore in Newfoundland, see Chapter 3.
4. On the use of vignettes in survey research, see Finch (1986).
5. There is a considerable history of spousal intermittence in Newfoundland, through employment in such activities as the trawler fishery, seasonal

Labrador fishery, merchant marine, forestry, and major construction projects. During this research, oil industry representatives of both the funding agency (ESRF), and the Hibernia project environmental assessment team argued that responses of Newfoundlanders to offshore oil work would be similar to those of other forms of employment. The clear rig/supply boat differentiation suggests that this is not the case.

6. While the rotation is longer, the pattern of work sees the supply boats in harbour on a regular basis, allowing the opportunity for contacts between workers and their families and between the spouses of workers (see p.176).

Afterword

Robert Moore

One important effect of oil development on women may be seen in the labour market statistics.[1] The regions in which the industry has located its onshore headquarters and its offshore services have been regions in which there has been a considerable growth of employment for both men and women. While for men the growth has been modest, and has largely had the effect of offsetting rising unemployment and the decline of traditional occupations, for women the growth has been absolute. *New jobs* have been created for women, whose economic activity rates have increased.

Thus while the total Norwegian work-force grew by 19.5 per cent from 1970 to 1980, the work-force in Nord Jaerun (three urban districts, including Stavanger) rose by 35.6 per cent. In the Aberdeen Labour Market Area employment grew by 40 per cent between 1971 and 1981 – the latter being the largest absolute growth of employment in the United Kingdom. The rise in *full-time* employment for men and women is shown in Table A.1.

The increase in female employment has been especially marked in Norway where traditionally the employment rate for women has

Table A.1 Full-time employment for men and women in the United Kingdom and Norway, the Aberdeen labour market area (ALMA) and Nord Jaerun, 1970/71–1980/81

	1971–81		*1970–80*	
	ALMA	*UK*	*Nord Jaerun*	*Norway*
	%		%	
Men	15	–7	4	–8.4
Women	13.8	3	21.9	12.7

Sources: United Kingdom Census 1971 and 1981, Norwegian Census 1970 and 1980, Bonney (1986).

been low. It appears that the 1970s was the decade in which Norwegian women moved out of the home into the labour market, with the female participation rate rising from under 30 per cent at the beginning of the decade to 40 per cent at the end. British female employment rose by only 4 per cent from 1971 to 1981, but from the higher base of 53 per cent. Not all of the growth, however, is in full-time employment. It seems that Norwegian women with higher educational qualifications moved into high status full-time jobs and the rest into lower status *part-time* jobs. But what is especially noticeable in the case of Nord Jaerun is that both *male* and female part-time employment increased threefold. This increase for men is unique to the three areas we are considering and demands further research; is it, for example, the result of Norwegian men becoming more home and leisure centred, preferring part-time work, or a particular effect of the development of the Norwegian economy forcing them into part-time work like women?

These changes in levels of part-time work in the oil-affected areas from 1970/71–1980/81 are shown in Table A.2:

Table A.2 Changes in full and part-time work for men and women in ALMA and Nord Jaerun, 1970/71–1980/81

		ALMA		*Nord Jaerun*	
		Men	*Women*	*Men*	*Women*
1970/71	F/T	63975	26728	33118	12857
	P/T	2189	15035	2759	5971
1980/81	F/T	73691	30818	34433	15671*
	P/T	1950	21636	8563	15534

Notes: * During the period the Nord Jaerun female population rose by 15.3 per cent compared with 5.6 per cent for the Norwegian female population.

The extent of the changes may be seen more clearly by indexing each cell in Table A.2 to 1970/71 = 100 (see Table A.3).

Part-time employment for Aberdeen women has, therefore, risen in the period by about 44 per cent. Thus of the 10 000 new jobs for women in ALMA, two-thirds are part-time. Parts of the City of Aberdeen have some of the highest female economic activity rates

Table A.3 Changes in full and part-time work for men and women in ALMA and Nord Jaerun, 1970/71–1980/81: 1970/71 = 100

		ALMA		*Nord Jaerun*	
		Men	*Women*	*Men*	*Women*
1970/71	F/T	100	100	100	100
	P/T	100	100	100	100
1980/81	F/T	115	115	104	121
	P/T	89	144	310	260

and the highest rates of part-time employment in the United Kingdom.

Most of the growth of employment in Aberdeen has been in services, retailing, administration and transport. Technical and administrative occupations have also grown rapidly for women in North Jaerun. Rosenlund (personal communication) has calculated, however, that low status, part-time work for women has expanded most rapidly, increasing over two and a half times while full-time low status jobs increased only slightly. For men such low status full-time jobs have declined but this was more than offset by an increase in part-time jobs of a similar status. Local employment growth in oil-affected areas offsets the impact of generally falling levels of employment in the United Kingdom and Norway. Thus while full-time work has declined by 2.7 per cent in Norway it has increased by 9 per cent for the population of North Jaerun.

The situation in Newfoundland is similar to both Norway and the United Kingdom but it would be rash to attribute changes to the impact of offshore oil developments. Activity off the Newfoundland coast has been slight, both absolutely and in comparison with the North Sea. The province has the lowest female work-force participation rate in Canada and marked rural-urban differences. This latter difference should perhaps alert us to the danger of disregarding female labour in rural activities not formally defined as 'employment'. The change in *all* employment for men and women in the province and St John's between 1971 and 1981 is as shown in Table A.4.

Women have risen from being 28.3 per cent of the work-force in the province in 1971 to 36.5 per cent in 1981 and in the same period

Table A.4 Changes in employment for men and women, Newfoundland and St John's, 1971–81

	Newfoundland %	*St John's* %
Men	20	21
Women	42.9	38

Source: Statistics Canada.

from 36.4 per cent to 42.2 per cent in St John's. There are then good reasons for being very cautious in attributing change to the impact of oil; Canada, Newfoundland and St John's have all had increases in the labour force, and women's activity has increased to an even greater extent. Yet neither Newfoundland, nor St John's were extensively effected by offshore oil in the 1970s.

The St John's area shows similar changes to Norway and Britain in the growth of part-time work, the magnititude of the changes are again made clear by indexing 1971 to 100.

Table A.5 Changes in employment for men and women, St John's, 1971 and 1981

		Number		*Index 1971 = 100*	
		Men	*Women*	*Men*	*Women*
1971	Full-time	27270	12940	100	100
	Part-time	2780	3285	100	100
1981	Full-time	28115	23400	103	181
	Part-time	4475	7935	161	242

Source: Canadian Labour Force Survey.

This book has not been concerned with an analysis of the conditions of women's work in the onshore oil industry. There is scope for considerable discussion and research, both of which we may hope will flow from the interest in women and oil raised in St John's, Newfoundland, in the autumn of 1985. There is need to study women onshore throughout the occupation scale. Teresa Turner's work in the onshore oil fields of Alberta is indicative of what needs to be done (personal correspondence). Turner also examined the impact of oil on native Canadian women, a group that has no exact equivalent in Scotland or Western Norway. In general terms, she found women working as contract staff in the labour camps under very unfavourable circumstances. The women had to cope with a transitting work-force of men, working in relative geographical isolation. They had to deal also with employer exploitation and with abuse and bullying from men.

Turner was, however, studying the oil fields, not the bases from which the oil industry was serviced and administered. In Alberta as in Aberdeen, nonetheless, it seemed much more preferable to work for an oil company than for a contractor. But, as I have suggested elsewhere (Moore, 1982, p.182), oil companies may prefer to subcontract certain kinds of work because the labour relations typical of that activity may not be in keeping with the corporate image of the company. The whole issue of subcontracting, as Heen signals in this volume, is of immense importance in understanding contemporary industrial developments.

One interesting finding in Chapters 4 and 6 is that the wives of men who are actually involved in offshore work are much less likely to be employed onshore. This raises questions about work-family relations that are the focus of the second half of this Afterword.

We might expect to find onshore oil offering many opportunities for women to achieve advancement in management and higher professional jobs. Hakim (1978), whose work is considered in more detail below, sounds a warning note, but her data refer to the United Kingdom until 1970. Important recent work by Crompton and her colleagues has suggested that where there is rapid expansion in a service sector, women, who are now graduating from higher education in increasing numbers, are likely to make substantial inroads into employment and promotion (Crompton, 1986).

She cites especially the rapid growth of the activities of the British building societies in the 1970s as just such a growth area. Here women did not have to compete with men and, as it were, win promotion against the men's interests. There were simply plenty of jobs and opportunities for promotion with well-qualified women available to take them (Crompton, personal communication).

Building societies, however, tend to recruit locally. Oil companies recruit nationally and internationally. They might not recruit heavily in Aberdeen, Nord Jaerun or St John's. None the less in Aberdeen we are aware of the pressures upon North American companies created by equal rights legislation at home. British companies too recognise the need both to protect their image and to cope with a demographic downturn in the late 1980s. Do we, then, find women moving into higher managerial posts in the oil industry onshore? The answer for Nord Jaerun seems to be 'Yes, probably'. Aberdeen remains to be researched.

One firm conclusion may, however, be drawn for Norway and the United Kingdom: the development of offshore oil fields has resulted in more women being employed onshore and has therefore raised incomes for women. This is unconditionally the case although the data so far presented hint at the inequality of men's and women's incomes and terms of service. These inequalities and explanations for them are the subject of much of what follows.

Offshore we have a different story to tell; only in the Norwegian sector of the North Sea do we find a significant number of women working offshore, and even here their geographical location is patchy. The Newfoundland Petroleum Directorate study found 2 per cent of the work-force to be female and forty of the 9311 workers registered for work offshore (see Chapter 3).

Whether we look onshore or offshore in Norway, the United Kingdom or Canada the story of womens' employment is the same. There is a very high degree of job segregation. Women work in different jobs from men, their work is very largely the servicing of men, and they receive lower pay than men. As Moore and Wybrow (1984) have shown, in 1983 3.8 per cent of the offshore work-force in the Norwegian sector of the North Sea was female but they constituted 37.5 per cent of the offshore catering work-force. They comprised only 1.1 per cent of the offshore work-force not engaged in catering. None the less, this is high representation when compared with British women offshore, who in 1984 comprised something like 0.1 per cent of the total work-force.

Contrary to popular expectations women are, in the United Kingdom at least, becoming *more* not less segregated in their occupations. Hakim (1978) examined data from the 1911 and 1971 Censuses and concluded that

> [the] likelihood of working in an occupation where one's own sex was overwhelmingly dominant became proportionately greater for men over the past seventy years. Male inroads into women's preserves have not been counterbalanced by womens' entry into typically male spheres of work.
>
> (Hakim, 1978, pp.1265–1266)

She goes on to say that her results

> confirm the pattern of women's under-representation in typically male jobs being much more marked than women's concentration in typically female jobs, even after taking account of the fact that men outnumber women 2 to 1 in the labour force.
>
> (Hakim, 1978, p.1266)

Furthermore

> they show very clearly the trend for women to become over-represented in the lower grades of work and under-represented in the higher grades . . . women were more evenly represented among managers and administrators in 1911 than at any time since then.
>
> (Hakim, 1978, p.1267)

When we observe this phenomenon of segregation to be so widespread and persistent, and so entirely non-random, we are bound to ask, 'how are these arrangements preserved?' and then, 'Whose interests do these arrangements serve?' These look like simple questions that probably have simple answers. But the questions are not simple and most of the answers are somewhat circular in form.

The interests served could be either 'ideal' or 'material' and the one may serve the other. Beliefs about the proper role of women vary from the inchoate fear and aggression, shown by some of the respondents in the Moore and Wybrow study, through to well worked out formulations. The former responses probably belong to the realm of the psychoanalyst. But the latter have an important place in the history of ideas and in our understanding of the construction of female roles. The ideas are especially important when women themselves subscribe to them, for the task of subordination is made simpler if the subordinated conspire in their own domination.

Ideas, embodied in practices, are part of the pre-market factors which constrain women's opportunities in the labour market. To contemporary readers past ideas about women make entertaining reading, like those of the nineteenth-century gentlemen and medical experts who believed that higher education would impair the reproductive capacity of young middle-class women. Interestingly enough, factory work did not seem to impair the capacity of working-class girls – but *that*, in itself, was a problem, to be targetted by the eugenicists within the early family planning movement, who were concerned with the decline of the national 'stock'.

But 'science' can still legitimise unequal treatment of men and women, with references to the hormonal control of female behaviour, measures of apparent differences in reasoning and motor skills or the dominance of brain hemispheres (Rose, Kamin and Lewontin, 1984, ch. 6). More popular accounts make the fact of childbearing and lactation the basis for unequal treatment or an imputation of female weakness. The fact of biological differences (which are few in number) has to be set in an institutional framework; bearing a child does not mean that one has to wipe its nose throughout its dependent years. Patterns of child care are features of *culture* and we need to analyse the role of biological differences as mediated by cultural definitions and structures. It is our particular *social* arrangements that make childbearing the basis for unequal treatment in the labour market, not the biological fact.

Women may now be more actively seeking to make their own history, but they do not make it in circumstances of their own choosing. Thus, for example, the education and training of women has not fitted them to compete in the labour market on equal terms with men. Historically education has been geared to assumptions about gender roles. This is clearly the case in the United Kingdom, Norway and North America. The (English) Board of Education reviewed the education of women in domestic skills at the beginning of this century. Girls were to learn sewing and cookery. An important issue in developing this education was what to do with the boys while the girls were at these classes. One solution was to have the boys make sewing boxes for the girls, or to study current affairs (civic housekeeping). Only in Norway were cookery schools established for boys – to meet the need for sea-going cooks (Great Britain, Board of Education, 1906, p.223).[2]

The past success of the education system in channelling women towards particular roles can be seen in the statistics of education, in school leaving ages, in qualifications acquired and in subject choices in higher education. In interviewing students and graduates Moore and Wybrow encountered cases of women who, as school girls, were either discouraged or prevented from doing science by their parents or their school. Also disadvantaged were those young women whose schools, in the course of educational reorganisation, were merged with boys' schools. The boys had 'squatters' rights' in the laboratories and in the interest of balanced class sizes girls were steered away from science.

But yesterday's taken-for-granted assumptions may become today's problems, as evidenced by the efforts now being made to encourage young British women into science and engineering and the recognition by employers that scientific and technical jobs need to be made attractive to women if employers are to maintain skilled work-forces through a period of demographic decline.

Inequality is not sustained by pre-market or ideological factors alone. Men may organise directly to keep women out of certain occupations through their trades unions and through legislation, or they may individually use violence to keep women in 'their place'.

The role of the trades unions in Britain in the nineteenth century is well-known. They sought legislation to protect women from conditions in the mines and factories and thereby protected their own market position. This was a strategy that worked in an age when interference with market forces was thought to be unrealistic or impossible. Women and children needed a special status in men's eyes to be so protected. A consequence of this was the reinforcement of women's dependent status. But if another consequence was the enhancement of men's market position, or limitations on their hours of work, then it was 'unintended' and not a tampering with the laws of nature. In the United Kingdom, legislation to limit the hours of male workers was not passed until the twentieth century.

If nature made woman the weaker vessel, the laws of nature were not entirely inflexible, none the less when women were allowed to undertake 'men's work' in the two world wars it was on condition that the men took their jobs back after the war. 'Dilution of labour', whether by women or migrants, was a fear with an empirical foundation. But the problem of what it was about women's status that made them potential dilutees was not addressed by the unions for whom post-war 'business as usual' meant back to work, supporting a wife and children at home.

Employers, too, may make a positive use of gender. Fevre's work has shown how women may be used in, for example, achieving a technical transition in the textile industry and how a similar role may be taken on by Pakistani immigrants (Fevre, 1984). In these cases employers manipulate gender- or ethnically-based statuses for particular purposes in a calculative way. Such labour *may* be 'cheap labour', employed as if it was not in a market at all. But far from groups being singled out as cheap labour, Fevre argues, certain poorly paid jobs continue to exist only because there are groups of the appropriate status to do them. More importantly perhaps female or immigrant labour may be employed as 'green' labour, unaccustomed to the traditional ways of an industry and thus willing to work in a manner that breaks with the customs and practices of the industry. This means that people of subordinate status are used to overcome the organised interests of an established work-force, to reduce worker autonomy and enhance management control.

That these are active processes and not historical residues is demonstrated by the British experience of the Equal Pay Act. Employers were given five years in which to prepare for the implementation of this legislation. In the five-year period, steps were taken to exclude women from its effects by the redefinition or segregation of jobs so that women were not doing the 'same' work as men. That women are now pursuing equal pay for work of equal value is in part a mark of the success of the earlier management strategy but also a recognition that women may do work that no men do and which is of high value to the employer, but paid much less than men. The first equal value case was that of a skilled industrial cook who claimed equal pay with skilled men in the shipyard where she worked. It is, however, important to recognise that men's definitions of women's place have never been entirely uncontested.

HOW DIFFERENT IS THE OFFSHORE OIL INDUSTRY?

In part, the answer to this question depends upon the explanation given for women's universally low status in the labour markets of the industrialised world. It is here that the circularity of argument may become most evident: explanations are couched either in terms of the working of market forces or the results of male dominance – neither explains why the market works against women nor how males dominate, and the evidence for either is the outcome one wishes to explain.

There are explanations which emphasise the advantages to employers of a cheap and unskilled workforce (Beechey, 1983). Braverman (1974) and others have also emphasised the role of women in 'deskilling' work. According to such analyses equality of opportunity and equal pay would undermine the economic value to the employer of female labour. But this argument cannot be wholly adequate. We should note that work may be deskilled *and then* 'feminised', as when the typist replaced the clerk at the beginning of this century after the introduction of the typewriter. Also there is no necessary connection between feminisation and deskilling. Coal mining, for example, has been to a large extent deskilled, but it is still a male preserve (Thompson, 1983, ch.7). But, setting these reservations to one side, two important questions remain about the simple attribution of women as cheap labour and agents of deskilling.

First, why is it possible to employ women as cheap labour and in a deskilling role? Secondly, if women are simply cheap labour, why have they not entirely replaced men in the labour force? Male interests are plainly organised to prevent their replacement by women but important ideological factors may also be at work. For example, the offshore installation is seen as 'a man's world', although this view is slightly less strongly held in Norway than elsewhere. There is a long 'macho' tradition in oil. This tradition also contributes to the exploitation of men in providing an environment in which they may be pushed beyond their own physical limits or the limits of safety. Masculinity emphasises the idea of man against the elements at the frontier of technology, rather than men collectively against the well-organised interests of employers.

We have one example of women being used in what looks like deskilling in the oil industry:

> Career openings for women in petroleum-related geo-sciences only appeared with the increased application of a scientific technology to oil in the twentieth century . . . One geologist who did some subsurface work recalled that this development opened real career opportunities for women because 'a lot of men don't like to sit at a desk and do tedious work'.
>
> (Olien and Olien, 1982, p. 90)

But here women were not used to deskill: they were, in fact, in the forefront of new scientific developments. But they were employed to do tedious work while the men worked in the field on 'real men's work'. So in this case, the idea that women are better suited to boring jobs saves men from the tedious work – a process that does not deskill men, but rather women.

A rather different argument, based on the idea of the existence of a dual labour market, was developed in the United States to explain differential white and black employment experience. It has subsequently been applied in the United States and the United Kingdom to the employment of women. In the latter case, the application has not been supported by much substantial empirical research.

Barron and Norris argued that the labour market could be treated as divided into two sectors. Workers in the primary sector have more permanent employment and training in skills that may be transferable; their jobs are better paid and more interesting with prospects of a career through a measure of upward promotion through a firm, or by seeking promotion between firms. In the secondary sector there is less security of employment, little opportunity to acquire skills and movement is horizontal between unskilled or semi-skilled employment. This is a good description of work in an offshore catering crew. Furthermore, secondary sector employment may offer none of the fringe benefits and pension schemes enjoyed by the worker in the primary sector. Also the work itself is more likely to be uninteresting with no intrinsic rewards and lower pay. The degree of permeability between the segments is not specified, but must be relatively low if the market is to be usefully regarded as dual.

Who are the secondary workers? Barron and Norris set out five attributes that are likely to make a group a source of secondary labour.

1. They are easily dispensible.
2. They can be differentiated from workers in the primary labour market by a conventional social difference.
3. They have a relatively low inclination to acquire training and skill.
4. They do not rate economic rewards highly.
5. They are relatively unlikely to develop solidaristic relations with fellow workers.

The idea of the dual labour market has advantages; the approach treats inequality as deriving from managerial strategies rather than the play of simple market forces. It reminds us of the importance of the different types of employer: the multinational may be the 'primary' employer and its subcontractors, or smaller down-market firms in general, the employers of secondary labour. This analysis breaks down, however, when we consider the worldwide strategies of companies in technically advanced fields that seek out 'secondary' labour in the Third World. But as Beechey (Kuhn and Wolpe, 1978, p.176) points out, to suggest that women possess the attributes listed relies heavily upon stereotypes and also undermines the notion that women's position can be explained in terms internal to the labour market. Jane Kenrick goes further:

> What is striking here is the similarity between arguments used to explain which workers occupy which sectors and classical statements justifying discrimination against women in the workplace in terms of their low commitment to work.
>
> (Kenrick, in Wilkinson, 1981, p.169)

In the dual labour market analysis the secondary sector is really a residual category. The main value of the model may be in understanding the primary labour market and its advantages to employers and a section of the employed.

There are, however, some economic arguments which suggest that discrimination does derive from market forces. The idea of *human capital* has commonly been used: human capital is acquired through education and training, including training on the job. The individual will be able to command rewards according to their investment. Women have expectations which lead to low investment and their capital depreciates as a result of absences from the labour market.

> Forgone market-oriented human capital of mothers is a part of the price of acquiring human capital in children, and, more generally, a price exacted by family life.
>
> (Caroline Freeman quoting Mincer and Polachek, in Wilkinson, 1981, p.135)

Women, as we know, are heavily represented among the lower paid, and, according to human capital theory, low-wage labour is 'often badly paid, not because it gets less than it is worth, but because it is worth so appallingly little' (Paul Ryan quoting Hicks, in Wilkinson, 1978, p.6).

This model plainly leaves much to be desired. If one accepts that *worth* is a factor in income it is highly misleading to suggest that it derives from a series of individual choices. The state, for example, 'invests' in labour educaton and households generate motivations and constraints on labour market behaviour. But no doubt these constraints and commitments could be quantified as 'utilities' and the outcome of choices (however arrived at) would be varying units of human capital. Robert Buchele, however, suggests that with one exception studies have found the difference in human capital variables explains less than 10 per cent of the earning gap between men and women. 'The rest is attributable to "discrimination"' (Wilkinson, 1978, p.212).

If human capital explains so little and the secondary sector in dual labour market theory is simply a residual category of non-primary workers, then it may be possible to think of women as part of a 'reserve army'. This notion derives from Marx, who wanted to see despecialisation for factory and mill wage-slaves, who could then become 'critical critics', and so on. But nowhere do we find him saying that the mother should despecialise in order to be a mother in the morning, shepherdess in the afternoon and critical critic in the evening. The problem of child and husband care is taken for granted as the individual woman's work and alternative provisions of care are not discussed, and have seldom been discussed in Marxist writing.

Irene Bruegel has suggested that the increased involvement of women in wage labour fits the picture Marx drew of an expanding reserve army:

> As capital accumulated, it threw certain workers out of employment into a reserve army; conversely in order to accumulate capital, capital needed a reserve army of labour. Without such a reserve, capital accumulation would cause wages to rise, and the process of accumulation would itself be threatened as surplus value was squeezed . . . [Marx] was concerned to show how the expansion of capitalism inevitably drew more and more people into a labour reserve of potential, marginal and transitory employment.
>
> (Bruegel, 1979, p. 12)

Bruegel concludes that the nearest we come to finding a true 'reserve army' is in part-time work in the service sector – precisely the area of maximum growth of employment for women, including women affected by the oil industry.

The Newfoundland provincial government study said that catering workers were the 'lowest paid, least satisfied, have the least job security and have less opportunity for advancement of anyone on the rig' (Fuchs *et al.*, 1983, p.86). Women compete for these jobs in a large pool of unskilled labour. While women may find themselves in such a 'reserve army' and most men may not, there are men with the women in the reserve none the less.

Bruegel also provides us with some resolution of the questions of *either* why dominant males did not keep women out of the work-force, or at least make them bear the brunt of rising unemployment, *or* why women do not replace men because they are cheap labour: women have been losing jobs and suffering discrimination in the areas where men have been traditionally dominant. These are the declining industries in which men are able to exploit the weak market position of women in order to defend their remaining interests. This loss of work by women has been more than compensated for by the post-war growth of women's employment in services. The services are less vulnerable to recession than the traditional male areas of employment. Thus two apparently contradictory processes – women losing work before men, women substituting for men – take place at the same time. But the areas in which female employment has grown may begin to decline as the market advantages to the employer of employing women decline. High technology may replace women workers in offices, banks and shops and

> Hence many groups of women who have traditionally regarded their jobs as secure will find themselves threatened with rationalisation on a scale comparable to the wholesale elimination of jobs in the traditional male strongholds – mining, railways, docks.
> (Bruegel, 1979, p.20)

The idea of *a* labour market for women may be far too simple. Explanations of women's economic status may not be easily read off from existing formulations, we could do no better than start from Philips and Taylor's cautionary comment: 'The sexual division of labour in wage work cannot be seen either as a product of patriarchal imperative on the one hand, or the long march of capital on the other' (Thompson, 1983, p.208).

So far I have presented a few facts about women in the oil industry and reviewed a range of theories about women in the labour market. Is this simply an academic exercise? Different arguments might lead to different strategies for remedial action. Are women's low employment status and low wages the result of simple prejudice that will yield to education and declaratory legislation; do they derive from vested interests embodied in restrictive practices that demand legally enforceable remedies; or is women's subordination so intrinsic to the social structures and the functioning of industrial societies that only a radical or revolutionary alteration of those structures will suffice?

If, for example, women are segregated because of their low human capital then employers can have no objection to the notion of equal pay for work of equal value. Women will then be able to receive the appropriate return on their investments. But economic interest and market forces cannot fully explain segregation and discrimination in the labour market. An economist warns us

> the power relations of capitalist society run deeper than the market forces that ostensibly determine wage rates in the capitalist economy. It is important to recognise that the hierarchical relations of production that shape our society's very definition of value, as well as its distribution of income, will not yield easily to redefinitions of worth that radically alter that distribution of income.
>
> (Buchele, in Wilkinson, 1978, p.224).

Employers may have 'utilities' of an ideological nature that modify the expected outcomes of the operation of market forces. Caroline Freeman (Wilkinson, 1978) reported the case of an employer desperate for labour, scouring a region and providing transport for female workers. This employer provided youth clubs, sports fields and other benefits for his workers, but not a crèche – potentially the most important single factor in the recruitment of women workers. The employer's *beliefs* about women counted for more than his economic interest. This may also be the case in the offshore oil industry. According to Clark and Taylor (Chapter 4), employers are entirely indifferent to issues relating the family and work. They 'employ men not families'. Not only is this untrue in all but the narrow contractual sense, but it may be contrary to the employers' own interests to believe and act as if was true. Perhaps then the task for women is mainly one of changing the minds of employers and to make the employer's self-interest the main thrust of persuasion?

WORK/FAMILY RELATIONS

In considering the role of women in the labour market we have seen how assumptions about women's other roles are made and used. 'The home' is the main location of female work, here they do heavy and continuous work, without a wage. The woman's work is the reproduction of labour both in the sense of producing the next generation of workers and in sustaining the current work-force – repairing and maintaining its physical and mental fabric. Oil men, discussing the employment of women offshore, raise the question of pregnant women – could they do the heavy work? They seemed largely oblivious of the heavy nature of the work undertaken by women, pregnant or not, in running a household with children and providing a home for a working man. Offshore work, with its regular shifts, might be easy by comparison.

British social policy seems to be geared not only to forcing women out of the labour market and back into the home, but increasingly to shifting caring work to the household. 'Community care' in Britain usually means unpaid care by women. The European Court has ruled against the policy by which Britain pays men allowances for caring for dependants (they need to buy in help) but not women (it is their normal duty). Much care of the elderly, for example, falls to women, who may have to give up work and suffer substantial loss of income.

Child care is *par excellence* the role of the mother. Looking after young children takes women out of the labour market, isolates them and ties them to the repetitive and unavoidable routines of domestic work. Even part-time work is difficult for women with school-age children unless employers are very flexible about hours of work and problems arising from illness and school holidays. For many working-class mothers, especially those without male support, the only option is to leave children locked out of the house or in the vague care of neighbours or older siblings. In one of the areas of high part-time female employment in Aberdeen, it is normal to find children under the age of ten locked out of their homes after dark.

Where limited child care is available it is, for younger children, the provision of 'mother substitutes' and for older children 'education'. There is in the United Kingdom no provision for, or proposal to provide, collective care for children. Only when 'normal' mothering is not available (i.e. female alone with children during working hours) may some essentially secondary help be provided.

According to Barrett and McIntosh: 'The imagery of idealised family life permeates the fabric of social existence and provides a highly significant, dominant and unifying, complex of social meaning' (Barrett and McIntosh, 1982, p.29).

Far from family life being a separate and private sphere the ideal values of a particular form of family life permeate other institutions, including the organisation of the labour market. Barrett and McIntosh locate the meanings associated with family life quite centrally in capitalist society:

> Just as the family has been socially constructed, so society has been familialised. Indeed it can be argued that in contemporary capitalist society one dominant set of social meanings is precisely an ideology of familialism. The meaning of family life extends far beyond the walls of concrete households.
>
> (Barrett and McIntosh, 1982, p.31)

They go on to illustrate this, beginning with the saturation of the mass media in familial ideology and the celebration of the ideology in the institution of the monarchy ('the royal family').

Ideology and reality may be at a remove from one another. A high proportion of marriages in industrial societies end in divorce. The divorcees then usually affirm their faith in marriage by remarrying. In Europe and North America children run away from home, living rough, even by prostitution, in order to escape the unbearable tensions of family life. Both the tensions and the isolation of family life make women vulnerable to the aggressive marketing of the pharmaceutical industry. Measured on both national and international scales the usage of psychotropic drugs is higher for women than for men – twice as high in the United Kingdom, where one in ten women have taken such drugs in any year. The results of taking these drugs is, for women, 'the perpetuation of their passivity and domesticity' according to Betsy Attore (Henman, 1985, pp.114–15). That this is the *intended* function of these drugs would seem to be borne out by the sales literature.

This analysis contrasts very sharply with traditional accounts of the family. Conventional accounts of industrialisation include the separation of the work-place from the home. The home is seen as a haven from the market-place and the factory, where affection and mutual support replace competitive market relations. A functional fit between the nuclear family and industrial society was found, they were complementary institutions. Parsons and Bales (1955) elaborated this with an account in which the division of labour in the market-place was complemented by the domestic division of roles. The man was the instrumental leader, he got things done and brought home the family income from the world of work. His spouse was the sociometric star; she held the family together, radiating love and warmth, providing sustenance and affection for the man worn down by work, and an environment suitable for raising children.

The image is well-known and the ideal celebrated in film, sermon, song and political rhetoric. It is far from the truth. Not least is it untrue because many women do two jobs, running the household and paid work. Furthermore, a growing proportion of families are headed by women and in the United States more than a third of these families live below the poverty line. But the ideal is one that many aspire to. Marriage is popular. In spite of what newspapers say, sociologists are not predicting the imminent demise of marriage. The most enlightened family policy is geared to preserving this family form and to easing the burdens of family life, especially for women.

To outsiders Norway seems an especially family-oriented society, with governments that have been prepared to back rhetoric with resources on a scale unknown in the United Kingdom and North America. These latter countries have governments who are generous with their familialist rhetoric but whose policies have made family life impossible for many. Familial rhetoric becomes oppressive by stigmatising those who do not fulfil its ideals and blaming them for many of society's alleged ills. It also stigmatises women who are in waged work and makes subordination and low pay easier to enforce while undermining women's image of themselves in terms of an idealised womanhood.

Political disruption of family life is not accidental. Recent changes in British social security regulations keep young people on the move, seeking jobs from place to place and losing entitlements if they stay too long in one place. The idea is to promote mobility of labour and discourage youths claiming their benefits in seaside resorts.[3]

Here we come close to the nub of the matter, not only is family structure geared to the labour market, family policy may also be geared to the need to get family members into or out of the labour market, or to using their labour to provide a private welfare state in default of adequate collective provision. Similarly familial ideology makes it possible to consign women to various kinds of subordinate roles in the market-place. The notion that women do not *really* belong in the sphere of waged work, but that their work is basically in a servicing function derives from strongly held ideas about the family.

There is, therefore, potentially of a problem with those papers in this collection that are primarily concerned with the adjustment between family life and work, the adaptation of families to work offshore and the need to facilitate easy transitions from home to work, in that they risk underwriting the very social forces holding women in economic subordination. But the more general conclusion to be drawn from this collection is that if there is a problem, it is not the family, but the economic and political circumstances of family life.

Offshore work emphasises many of the contradictions of life in the nuclear family. Men do, indeed, see the family as a haven from the world of work, as Parsons suggested. But the men expect the family to make the adjustments on their return from work.

Thus an extra burden is created for women, and especially if they have full-time jobs. In Aberdeen and Norway women reported how men expected them to drop their obligations when they came home. Thus while the man's absence may increase women's independence in the labour market this independence and other autonomous relations that women develop while he is away come under particular stress. The increased independence of women, however, is something that men find hard to adjust to (see Chapters 5 and 6).

The more traditional the man's view of marriage the more difficult it is for the woman to adjust. In the Norwegian and Aberdeen (but not the Newfoundland) cases traditional families seem to handle the stress more effectively. But the wife is especially isolated when the man is away and the man needs to find something to do on his return. What men do – extending the house, building a boat – is an extension of 'man's work' in the labour market. Thus, just as work relations may be familialised so family roles reflect the labour market. This 'man's work' also reinforces the role of the instrumental leader in the family.

Solheim's evidence (Chapter 5) suggests that in Norway family-centredness makes offshore work *more* difficult to cope with. In the more companionate family, with shared roles, men have less of a transition period between work and home because they are expected to be an immediate participant in the life of the family on their return. Yet the conditions of work leave them less able to be so. The material benefits of work offshore seem however to compensate for the disruption, as evidenced by the responses reported in the Newfoundland research.

It seems as if the organisation of offshore work assumes the existence of a traditional family type. Conversely, given the problems of the transition from work to home and back again described in Part II it may be that this kind of work will constrain other types of family *towards* a more traditional structure.

We might look for evidence of family stress in statistics of separation or divorce, but this cannot be done by using figures based upon geographical areas, and oil workers' families are impossible to pick out of national statistical tabulations. Samples of oil families will be needed. Work of the kind undertaken by Clark and his colleagues shows us what can be done to locate a suitable sample for study. Moore's (1982) experience in Peterhead was that where marriages had ended in divorce or separation it was because the availability of work made it possible to escape marriages that had long broken down. All the contributors to this volume found little increase in the breakdown or destruction of marriage. None the less this observation says nothing about the quality of the relationships within the marriages that seem so robust.

The most interesting theoretical theme in the discussion of work and family life made explicitly by this book is its challenge to the traditional work-home dichotomy. Solheim, for example, noted in Chapter 5 that some women do not have a problem with men being away offshore. Their main problem was when the man came home and disrupted their routines. The division was not, then, between home and work but between offshore work, the woman alone, and family life. So the man was making a work-home transition while the woman was making a work-family transition. In other words, the man did not come home to something that was there when he was away, but to something that was created for his return and which constituted a very considerable burden for the woman.

An important and original theme in the papers on family life is the recognition of men's dependence on the family *and* their emotional needs and stresses. This may be lost sight of in conditions of normal work ashore where stressful work comes in shorter daily bursts and, more importantly, men have the opportunity to 'unwind' in the pub or elsewhere, with workmates before going home. This opportunity is, of course, denied working wives who have to go straight home from work to prepare a meal for their man who is unwinding in the pub. Alcohol plays its own part in family stress, but it also performs a valuable role in effecting the transition from work. It also raises questions about the nature of work that it demands the use of a drug to make it bearable.

The study of oil work sharpens our awareness of conflicts that are found throughout work and family relations. None of us are immune to feelings of tension when we make the transition from work to home, but the transition is a daily affair and may take a variety of forms during the week. We can all try to do better tomorrow. However, in the oil families on both sides of the Atlantic the tensions are exacerbated by the expectations that both parties have of their less routine reunions.

Another question thrown up by the analysis of differences in work culture is this: is it possible that particular work cultures afford more protection for workers than other cultures? One striking contrast between the offshore work environment in the British and Norwegian sectors of the North Sea is what a Norwegian observer described as 'the pub culture' of the British worker. Off-duty workers played cards or dominoes together, he said, or they took part in organised activities including raising money for onshore charities. This pattern of warm sociability was contrasted with the Norwegian worker sitting in his cabin thinking of home, or of his boat. The contrast may be slightly overdrawn but Heen's Chapter 2 notes the tendency to withdraw or become 'lonesome' among Norwegian men offshore, and while men interact mainly with their immediate workmates there is also a tendency to avoid relationships.

What the British see offshore is immediately identifiable as a rather traditional British male work culture. The warm sociability of the club and pub was a continuation and alternative expression of the relations of dependency that men experienced at the workplace. The strong bonds thus forged provided a basis for trade unionism and Labour politics. Male political consciousness was rooted in a triangle of work, union and club or chapel. Women lacked this double or treble bonding, save in a few industries where women worked together close to their homes. The decline of the traditional industries – shipbuilding, mining, railways – has seen a decline of the occupational community. Workers in growing industries tend to live apart from one another, mixed in with workers from other occupations and of various statuses. Their experiences are more like women's experience of work, and as we know, many of them *are* women.

The 'steel island' may provide a new type of occupational community, albeit one very different from past communities. The Norwegian studies suggest that this is so, and that women may be fully integrated into the community but less integrated perhaps into the work.

Male solidarity in the United Kingdom enabled men to develop the exclusionary practices that have already been described. Women had no such opportunities. One reason why men onshore seem to resent a female presence it that a woman is not 'one of the boys'. Women do not go drinking after work together, they are not part of the tight solidary group that men develop at work and play. The solidarity of the drill-floor work group is well described in Chapter 3. Solidarity has, for many male workers in the United Kingdom, been their only protection in ensuring acceptable wages and conditions, and securing safety at work. Acts of solidarity have, however, been made illegal in the United Kingdom since the passage of the 1980 Industrial Relations Act.

The question of work-group solidarity is plainly of considerable importance in considering safety offshore and offshore unionisation. These are not, however, the direct subject of the present discussion, but what is of particular interest is whether the lack of a 'pub culture' in offshore Norway leads to an accumulation of tensions that are expressed only when the workers return home.

Norwegian men approve of women working with them offshore and some British-based employers have said they have no objection in principle. Moore and Wybrow were told how this made life more 'normal'. What we were told was true. Men like women working as cooks, cleaners and secretaries offshore.

The nearest that men have to 'home' when they are offshore is their cabin and the canteen and recreational space. But the 'domestic' side of life has been defamilialised – there are men doing the cooking and cleaning. How much more normal and acceptable then for the offshore environment to be refamilialised with women doing their usual work? The comments that we heard about men taking more care about their personal appearance when women were on board suggests that women are still a 'civilising' influence of the kind envisaged by the domestic science teachers of the early twentieth century. They suggest also that it takes little for men to revert to a less civilised form of life without female support. In both Norway and Newfoundland women were reported as being under pressure to perform part of the role of 'sociometric star' in listening to men's personal problems.

Men did not, however, like women joining their own work-group and Heen has briefly described the conflicts and ambiguities of the role of a woman in a men's work-group. In the United Kingdom sector Moore and Wybrow met women who had worked *with* men offshore, but not *in* the work-group. These women stood in more of a professional role vis a vis the production work-force – their work-group was a small group of professional peers. But there were plainly conflicts for some of these women in relating to men with traditional attitudes, like the tool-pusher who said he had never taken orders from a woman and did not intend to start now. Fifteen per cent of the sample interviewed by Anger, Cake and Fuchs were hostile to women working offshore and 35 per cent were not. But being 'for' women offshore does not mean accepting women in one's own work-group, or being willing to take orders from women, any more than entering a marriage 'partnership' entails accepting a woman as a peer.

A second major conclusion to be drawn from the research presented in this book is that the family-work dichotomy, derived from functionalist theories of industrialisation is highly misleading. As David Clark and Rex Taylor stated (p.112):

> Such a distinction often serves to mask the complex interrelationships between work and the family and is likely to obscure important gender differences which shape exprience in the two arenas.

It may be convenient for the purposes of certain kinds of research to separate family life and work, but this heuristic distinction has permeated our thinking to the point where it has become a false reality. Nor can we make sense of the most elementary aspects of work – the division of labour, for example – without considering both the family and familial ideology.

CONCLUSION

Let us conclude by reviewing problems of strategy in three fields. First, can anything be done within existing political frameworks radically to alter the status of women in the labour market? Any answer to this must be equivocal. But women have nothing to lose by exploiting to the full the opportunities that *are* available, few though they may be. Norway, for example, has by far the most advanced policies for the support of family life, it has had equal pay legislation since 1959 and in 1977 introduced a potentially far reaching Act on Workers' Protection and Working Environment. The creation of social conditions that made such provisions possible represents the success of major liberalising forces in Norwegian society. Women have made progress, the laws which offer them further opportunities are monitored carefully and women will find fewer legal or political barriers to their progress than women in many countries.

Britain has a decaying and underfunded welfare state, sex-equality legislation is only weakly enforced: furthermore, it is subjected to continuous derision in the press and is not applicable offshore. Equal pay became part of Labour party policy only in 1947 and the Equal Pay Act was not passed until 1970. Major campaigns are still needed (and some have begun) to gain a place on the political agenda for the rights of women. Many of the limited post-war measures of family support need to be reclaimed or reconstructed. Membership of the European Community and obligations under the European Convention on Human Rights offer important channels for pursuing United Kingdom women's interests.

According to the International Labour Organisation, Canada has an affirmative action programme for employers, daycare is increasingly funded through public subsidy and all provinces regulate the operation of daycare facilities. Some provinces (including Newfoundland) make special efforts to help women move away from sex-stereotyped jobs, some unions operate daycare facilities for their members. The government of Newfoundland is committed to enabling women to benefit from oil developments and *publicly* restated that commitment during the 1985 'Women and Offshore Oil' conference. Words are not enough and Newfoundland women seem to have organised themselves to make their voices heard in claiming their rights to consultation in and benefit from oil development. To a British observer the degree of organisation both at community level and in formal bodies is impressive.

However radical the policies may be that are needed substantially to alter women's status in society, reformist policies are worth pursuing, among others. This much would be indicated by Peter Sloane, whose basic approach is that of a human capital theorist:

> Not only does [Sweden] possess the smallest earnings gap between men and women among the advanced industrialised nations, but also, and presumably not by accident, the narrowest differential between male and female labour force participation. This outcome has been reached not through legislation, a sex equality law only being introduced in 1980, but through social policies which have encouraged female labour force participation. These include the separate taxation of husband and wives' earnings . . . paid parental leave financed by general taxes . . . and abundant day care facilities.
>
> (Sloane, 1985, p.30)

Perhaps a radical transformation is needed to create the circumstances in which such reforms might be enacted.

Why should women not continue to pursue the channels that are open to them and grasp the opportunities that are offered? One important reason why they should is that only then will they be able to test the extent to which social structures will really yield to their interests. It might bring them to a stage in which more radical changes are seen to be needed.

Secondly, to what extent are women prepared to take advantage of the opportunities offered by offshore oil? 'Prepared' may be understood in two senses. First, do women really want to get into oil? The answer to this question would seem to be negative save for a few professional women and some catering workers in search of higher wages. But women's expectations are partly a product of their individual and collective experience. They do not see the point of seeking work in a 'man's world'. One can only make a very hard-nosed response to this; the oil companies and some other contracting companies say they need women in the industry, the Norwegian and Newfoundland governments want women to enter the industry. The industry must be tested. If 4 per cent of the offshore workforce and only 0.4 per cent of workers registered for work offshore are women it might otherwise be said that women are already heavily over-represented offshore.

But 'prepared' also means qualified. It is almost too late to change any of our education and training systems to meet the needs of women who wish to enter oil in skilled trades or professions. It is, however, not quite too late because production work will continue for many years in the offshore oilfields. Education and training, and perhaps compensatory retraining, are legitimate demands for all women and should be pursued.

Finally, there is a very awkward question. Can any of this make any difference so long as we adhere to a particular ideal of the family in which women occupy a servicing role for the work-force of this and the next generation? The pattern of job segregation corresponds closely to the domestic division of labour and much of the work that women do for wages is domestic labour performed outside the home. Domestic work at home is unwaged and therefore falls into the lowest possible category of low paid work. Women at work outside the household find their work similarly evaluated. Can women's position in the labour market be changed without a radical re-evaluation of their position in the household? In this context the early twentieth century attempts to professionalise the domestic role of women seem quite radical.

All the adjustments to work have to be made by women who bear a double or treble burden as a result. The nuclear family is immensely popular and many of the chapters in this book have celebrated that popularity and the resilience of the family.

Can women truly benefit from the development of the offshore oil industry or any other industry within the present arrangements for reproduction? We can be even more radical and ask, can industrial enterprises and the organisation of labour be reconstructed to create conditions in which new forms of 'family life' can be created which are not rooted in women's disadvantages? Anyone who has read the accounts of life working offshore is bound to conclude that such new forms would liberate men as well.

Notes

1. Cross-national statistical comparisons are notoriously difficult and within national statistical publications definitions may change between Censuses or surveys. Each nation, for example, has a different definition of part-time work and the boundaries of the St John's Census metropolitan area changed slightly between 1971 and 1981. No attempt has been made here to reconcile statistics from Norway, Britain and Canada, they are simply presented in their 'raw' form to indicate the direction of changes and the order of magnitude of these changes.

2. It is important to note what else this kind of education was doing. Women's work was being presented as professional work demanding skills and commanding esteem. Through domestic science attempts were also made to tackle problems arising from the straightened domestic circumstances of the poor, including those of hygiene and nutrition. Domestic *science* therefore had emancipatory aspects.
3. Very conveniently for the United Kingdom government, young people who keep on the move effectively lose their vote which is tied to the place of registration in February.

Bibliography

ACHESON, J.M. (1981) 'The Anthropology of Fishing', *Annual Review of Anthropology*, 10.

AMSDEN, A. (1980) *The Economics of Women's Work* (Harmondsworth: Penguin).

ANDERSON, R. and WADELL, C. (eds) (1972) *North Atlantic Fishermen* (St John's: Institute of Social and Economic Research).

ARDENER, Shirley (1975) *Perceiving Women* (London: Aldine Press).

BARRETT, M. (1980) *Women's Oppression Today* (London: Verso Press).

BARRETT, M. and McINTOSH, M. (1982) *The Anti-Social Family* (London: Verso Press).

BARRON, R.D. and NORRIS, G.M. (1976) 'Sexual Divisions and the Dual Labour Market', in D.L. Barker and S. Allen (eds) *Dependence and Exploitation in Work and Marriage* (London: Longman).

BEAUVOIR, S. de (1946) *The Second Sex* (Harmondsworth: Penguin).

BEECHEY, V. (1983) 'What's so Special about Women's Employment? A Review of some Recent Studies of Women's Paid Work', *Feminist Review*, no. 15.

BEECHEY, V. (1985) 'Conceptualising Part-time Work', in B. Roberts *et al.* (eds) *New Approaches to Economic Life* (Manchester: Manchester University Press).

BELL, Colin and NEWBY, Howard (1976) 'Husbands and Wives: the Dynamics of the Deferential Dialectic', in D. Leonard Barker and Sheila Allen, *Dependence and Exploitation in Work and Marriage* (London: Longmans).

BERGER, P.L. and KELLNER, H. (1964) 'Marriage and the Construction of Reality', *Diogenes*, no. 46.

BLAIR, J. (1976) *The Control of Oil* (New York: Pantheon).

BONNEY, N. (1985) 'Female Employment in the Aberdeen Labour Market Area, 1971–1984', *Women and Men in Scotland* (Manchester: Equal Opportunities Commission).

BORCHGREVINK, Tordiis and MELHUUS, Marit (1986) *Familie of arbeid: Fokus pa sjomannsfamilien* (Oslo: Work Research Institutes).

BOSS, P., McCUBBIN, H. and LESTER, G. (1979) 'The Corporate Executive Wife's Coping Patterns in Response to Routine Husband-Father Absence', *Family Process*, 18.

BRAVERMAN, H. (1974) *Labour and Monopoly Capital: The Degradation of Work in the Twentieth Century* (New York: Monthly Review Press).

BROWN, R. (1984) 'Work Past, Present and Future', in K. Thompson (ed.), *Work in Society* (London: Heinemann).

BRUEGEL, I. (1979) 'Women as a Reserve Army of Labour', *Feminist Review*, No. 3, 12–23.

BRUEGEL, I. (1983) 'Women's Employment, Legislation and the Labour Market', in J. Lewis (ed.), *Women's Welfare/Women's Rights* (London: Croom Helm).

BUJRA, Janet M. (1978) 'Introduction: female solidarity and the sexual division of labour', in Patricia Caplan and Janet M. Bujra, *Women United. Women Divided* (London: Tavistock).

BURGOYNE, J. and CLARK, D. (1984) *Making a Go of It. A Study of Stepfamilies in Sheffield* (London: Routledge & Kegan Paul).

CALLAN, Hilary and ARDENER, S. (eds) (1984) *The Incorporated Wife* (London: Croom Helm).

CANADA. (1984) *The Report of the Commission on Equality in Employment* (Ottawa: Queen's Printer).

CANADA, MINISTER OF SUPPLY AND SERVICES. (1985) *Canada Oil and Gas Lands Administration Annual Report, 1984* (Ottawa: Queen's Printer).

CARSON, W.G. (1982) *The Other Price of Britain's Oil: safety and control in the North Sea* (Oxford: Martin Robertson).

CLARK, D., McCANN, K., MORRICE, K. and TAYLOR, R. (1985) 'Work and Marriage in the Offshore Oil Industry', *International Journal of Social Economics*, vol. 12, no. 2.

COCKBURN, Cynthia (1983) *Brothers. Male Dominance and Technological Change* (London: Pluto).

COHEN, S. (1972) *Folk Devils and Moral Panics* (London: MacGibbon & Kee).

COMMUNITY RESOURCE SERVICES (1984) LTD. (1986) *Family Life Adaptations to Offshore Oil and Gas Employment* (Ottawa: Environmental Studies Revolving Fund).

CROMPTON, R. (1986) 'Credentials and Careers: Some Implications of the Increase in Professional Qualifications Amongst Women', *Sociology*, 20.

DAVID, Miriam (1983) 'The New Right, Sex, Education and Social Policy: Towards a New Moral Economy in Britain and the USA', in J. Lewis (ed.), *Women's Welfare/Women's Rights* (London: Croom Helm).

DAVID, Miriam (1986) 'Morality and Maternity: Towards a Better Union than the Moral Right's Family Policy', *Critical Social Policy*, no. 16.

DAVIDOFF, Leonore (1979) 'The Separation of Home and Work? Landladies and Lodgers in Nineteenth and Twentieth Century England', in S. Burman (ed.), *Fit Work for Women* (London: Croom Helm).

DAVIS, D. (1983) *Blood and Nerves: an ethnographic focus on menopause* (St John's: Institute of Social and Economic Research).

DEERE, C. (1976) 'Rural Women's Subsistence Production in the Capitalist Periphery', *The Review of Radical Political Economics*, 8, no. 1.

DELPHY, C. (1977) *The Main Enemy: a materialist analysis of women's oppression* (London: Women's Research and Resources Centre)

DENNIS, N., HENRIQUES, F.M. and SLAUGHTER, C. (1959) *Coal is Our Life* (London: Tavistock).

EICHENBAUM, L. and ORBACH, S. (1984) *Hva vil kvinnen* (Oslo: Cappelen).

FEVRE, Ralph (1984) *Cheap Labour and Racial Discrimination* (Aldershott: Gower).

FINCH, J. (1983) *Married to the Job* (London: Allen & Unwin).

FINCH, J. (1986) 'Research Note: the vignette technique in survey research', *Sociological Review*, (forthcoming).

FRANKLIN, U. (1985) *Will Technology Manage Women?* (Ottawa: Canadian Research Institute for the Advancement of Women).

FRIEDAN, Betty (1981) *The Second Stage* (London: Abacus).

FUCHS, R., CAKE, G., and WRIGHT, G. (1983) *The Steel Island: Rural Resident Participation in the Exploration Phase of the Oil and Gas Industry, Newfoundland and Labrador, 1981* (St John's: Department of Rural, Agricultural and Northern Development, Government of Newfoundland).

FUCHS, R. (1983) 'The Adaptation of Rural Residents to the Offshore Oil and Gas Labour Force, Newfoundland and Labrador, 1981', Paper presented to the Health and Welfare Canada Symposium on Psycho-Social Impacts of Resource Development in Canada: Research Strategies and Applications, Toronto, Ontario.

FURST, G. (1985) *Retratten fran mannsjobben*. Monograph from the Department of Sociology (Gothenburg: University of Gothenburg).

GAME, A. and PRINGLE, R. (1985) *Gender and Work* (London: Pluto).

GILLIGAN, C. (1984) *In a Different Voice* (Cambridge, Mass.: Harvard University Press).

GRAHAM, H. (1982) 'Coping or how Mothers are seen and not heard', in S. Friedman and E. Sarah (eds), *On the Problems of Men* (London: Women's Press).

GRAHAM, H. (1984) *Women Health and the Family* (Brighton: Wheatsheaf).

GRAMLING, Bob (1985) 'Socio-Economics Impacts of Offshore Oil and Gas Production: Generalising from Louisiana to Newfoundland' (Lafayette: University of Southwestern Louisiana, Unpublished Paper).

GREAT BRITAIN, BOARD OF EDUCATION (1906) Cd. 2963. 'Belgium, Sweden, Norway, Denmark, Switzerland and France'.

GREAT BRITAIN, DEPARTMENT OF ENERGY (1985) *Development of Oil and Gas Reserves in the UK, 1985* (London: HMSO).

GROSS, Harriet Engel (1980) 'Dual-Career Couples Who Live Apart: Two Types', *Journal of Marriage and the Family*, 42.

HAGEN, M. (1976) *Oljens Betydning for Kvinnelif Syssel-Setting i Rogaland* (The Impact of the Oil Industry on Women's Employment in Rogaland) (Oslo: Equal Status Council).

HAKIM, C. (1978) 'Sexual Divisions within the Labour Force: occupational segregation', *Department of Employment Gazette*, November.

HAKIM C. (1979) *Occupational Segregation*, Research Paper No. 9 (London: Dept. of Employment).

HAREVEN, T.K. (1982) *Family Time and Industrial Time* (Cambridge: Cambridge University).

HARTMANN, H. (1981) 'Kapitalism, patriarkat och konssegregationen i arbetet', *Kvinnovetenskapelig tidsskrift*, no, 1–2.

HEEN, H., ASLAKSEN, K., LIAAEN, O. SAGBERG, F. and QVALE, T. (1986) *Safety and Social Integration on the Ekofisk Field*, Report (Oslo: Work Research Institutes).

HELLESØY, O.H. (1981) *Health, Safety and Work Environment on the*

Statfjord Field. Stavanger: Mobil Exploration Norway Inc.

HELLESØY, Odd. H. Ed. (1984) *Arbeidsplass Statfjord. Arbeidsmiljo, helse og sikkerhet pa en oljeplattform i Nordsjoen* (Bergen: Universitetsforlaget).

HENMAN, A. (1985) *Big Deal: the Politics of the Illicit Drugs Business*. (London: Pluto Press).

HIBERNIA ENVIRONMENTAL ASSESSMENT PANEL (1976) *Report* (Ottawa: Federal Environmental Assessment Review Organization).

HOLLOWELL, P.G. (1968) *The Lorry Driver* (London: Routledge & Kegan Paul).

HOLTER, Øystein G. (1982) *Olje og skilsmisse* (Oslo: Work Research Institutes).

HOLTER, Øystein G. (1984) *Catering for Oil: Catering and the Reproduction of North Sea Communities (Oslo: Work Research Institutes)*.

HOUSE, J.D. (1985) *The Challenge of Oil: Newfoundland's quest for controlled development* (St John's: Institute of Social and Economic Research).

INTERNATIONAL LABOUR OFFICE (1978) *Employment of Women with Family Responsibilities* (Geneva: ILO).

KALUZYNSKA, E. (1980) 'Wiping the Floor with Theory', *Feminist Review*, no. 6.

KARLSEN, Jan Erik (1982) *Arbeidervern pa sokkelen* (Oslo: Universitetsforlaget).

KANTER, Rosabeth Moss (1977) *Work and Family in the United States* (New York: Russell Sage Foundation).

KØRBOL, A. (1970) *Kvinnen ombord*. En sosiologisk studie av kvinners liv og arbeid pa skip i den norske hardelsflate. Magistergradsavhandling. Institute for Sosiologi (Oslo: Universitet i Oslo).

KUHN, A. and WOLPE, A-M. (eds) (1978) *Feminism and Materialism* (London: Routledge & Kegan Paul).

LAND, H. (1980) 'The Family Wage', *Feminist Review*, no. 6.

LAND, H. (1983) 'Poverty and Gender: the distribution of resources within the family', in Muriel Brown, *The Structure of Disadvantage* (London: Heinemann).

LAND, H. (1986) 'Women and Children Last: reform of social security?' in Maria Brenton and Clare Ungerson (eds), *Yearbook of Social Policy* (London: Routledge & Kegan Paul).

LAXER, J. (1983) *Oil and Gas: Ottawa, the Provinces and the Petroleum Industry* (Toronto: James Lorimer).

LEIRA, A. (1978) *Kvinner pa en oljearbeidsplass*, Stavanger: Rogalandsforskning, Rapport, nr. 6.

LEVITAS, R. (1986) *The Ideology of the New Right* (Oxford: Polity Press).

LIAAEN, Ola (1982) *Verneombud, tillitsmenn og sikkerhet i oljevirksomheten* (Oslo: Work Research Institutes).

LIND, J. and MACKAY, G.A. (1980) *Norwegian Oil Policies* (London: Hurst & Co).

McCANN, K., CLARK, D., TAYLOR, T. and MORRICE, K. (1984) 'Telephone screening as a research technique', *Sociology*, vol. 18, no. 3.

MACINTOSH, H. (1968) 'Separation Problems in Military Wives',

American Journal of Psychology, 125.
MACKAY, G.A. (1984) 'Oil and the Oil-Related Sector', in N. Hood and S. Young (eds), *Industry, Policy and the Scottish Economy* (Edinburgh: Edinburgh University Press).
McNALLY, F. (1979) *Women for Hire* (London: Macmillan Press).
MARTIN, Jean and ROBERTS, Ceridwen. (1984) Department of Employment and OPCS. *Women and Employment. A Lifetime Perspective* (London: HMSO).
MARTIN, L. (1984) 'Women Workers in a Masculine Domain: Jobs and Gender in a Yukon Mine' in K. Lundy and B. Warme (eds), *Work in the Canadian Context*, 2nd ed (Toronto: Butterworth).
MOBIL EXPLORATION NORWAY INC. (n.d.) *Working Environment, Health and Safety on the Statfjord Field* (pamphlet).
MOBIL OIL CANADA LTD. (1985) *Hibernia Development Project, Environmental Impact System, Vol. II* (Toronto: Mobil).
MOLYNEUX, M. (1979) 'Beyond the Domestic Labour Debate', *New Left Review*, no. 112.
MOORE, R. (1974) *Pitmen, Preachers and Politics* (Cambridge: Cambridge University Press).
MOORE, R. (1982) *The Social Impact of Oil: The Case of Peterhead* (London: Routledge & Kegan Paul).
MOORE, R. and WYBROW, P. (1984) *Women in the North Sea Oil Industry* (Manchester: Equal Opportunities Commission).
MORGAN, D.H.J. (1975) *Social Theory and the Family* (London: Routledge & Kegan Paul).
MORRICE, J.K.W. (1981) 'Psychosocial Problems in the Oil Industry', *Update*, vol. 22.
MORRICE, J.K.W. and TAYLOR, R.C. (1978) 'Oil Wives and Intermittent Husbands', *British Journal of Psychiatry*, vol. 47.
MORRICE, J.K.W., TAYLOR, R.C., CLARK, D. and McCANN, K. (1985) 'Oil Wives and Intermittent Husbands', *British Journal of Psychiatry*, 147.
MOUNT, F. (1983) *The Subversive Family* (London: Allen & Unwin).
NORENG, O. (1980) *The Oil Industry and Government Strategy in the North Sea* (London: Croom Helm).
NORWAY, CENTRAL BUREAU OF STATISTICS (1983) *Labour Force Sample Surveys*.
NORWAY. Equal Status Council (1984) *Minifacts on Equal Rights* (Oslo: Equal Status Council),
NORWAY (1985) *Norwegian Equal Status Act with Comment* (Oslo: Ministry of Consumer Affairs and Government Administration).
OLIEN, M. and OLIEN, D. (1982) *Oil Booms: Social Change in Five Texas Towns* (Nebraska: University of Nebraska Press).
PAHL, J. (1980) 'Patterns of Money Management within Marriage', *Journal of Social Policy*, 9.
PAHL, J. (1985) 'The Allocation of Money within the Household', in M.D.A. Freeman (ed.), *State, Law and the Family* (London: Tavistock).
PAHL, R.E. (1984) *Divisions of Labour* (Oxford: Blackwell).
PARSONS, T. and BALES, R.F. (1955) *Family Socialization and*

Interaction Process, (New York: Free Press).
PATTERSON, J.M. and McCUBBIN, H.L. (1984) 'Gender Roles and Coping', *Journal of Marriage and the Family*, 46.
PORTER, M. (1983) *Home, Work and Class Consciousness* (Manchester: Manchester University Press).
PORTER, M. (1984) 'The Tangly Bunch: the Political Culture of Avalon Women', *Newfoundland Studies*, no. 1.
QVALE, Thoralf U. (1985) *Safety and Offshore Working Conditions. The Qualify of Work Life in the North Sea* (Oslo: Norwegian University Press).
RADCLIFFE-BROWN, A.R. (1952) *Structure and Function in Primitive Society* (London: Cohen and West).
RATHBONE, Eleanor (1936) 'Changes in Public Life', in Ray Strachey (ed.) *Our Freedom and its Results* (London: Gollancz).
ROBERTS, B., *et al.* (eds) (1985) *New Approaches to Economic Life*. (Manchester: Manchester University Press).
ROGNE, Karl, QVALE, Thoralf U. and OSTBY, Harald (1982) *Arbeidsorganisasjon pa produksjonsplattformer* (Oslo: Universitets-forlaget).
ROSE, S., KAMIN, L. and LEWONTIN, R. (1984) *Not in Our Genes* (Harmondsworth: Penguin Books).
SAMPSON, A. (1975) *The Seven Sisters* (New York: Viking).
SLOANE, P. (1985) *Sex at Work: equal pay and the 'Comparable Worth' controversy* (London: David Hume Institute).
SOLHEIM, Jorun (1983) 'Offshore Commuting and Family Adaptation in the Local Community' (Oslo: Work Research Institutes).
SOLHEIM, J. (1985) *Offshore Commuting and Family Adaptation in the Local Community*. Paper presented at the conference 'Women and Offshore Oil', St John's, Newfoundland (Oslo: Work Research Institutes).
SOLHEIM, Jorun and HANSSEN-BAUER, Jon (1983) *Complexity and Communality on a North Sea Platform* Report (Oslo: Work Research Institutes).
SPENDER, D. (1982) *Women of Ideas and What Men have done to them* (London: Routledge & Kegan Paul).
STACEY, M. (1981) 'The Division of Labour Revisited or Overcoming the Two Adams', in P. Abrams *et al.* (eds), *Development and Diversity in British Sociology 1950–1980* (London: Allen & Unwin).
TAYLOR, R.C., MORRICE, K., CLARK, D. and McCANN, K. (1985) 'The Psycho-social Consequences of Intermittent Husband Absence: an epidemiological study', *Social Science and Medicine*, vol. 20, no. 9.
THOMPSON, P. (1983) *The Nature of Work: An introduction on debates on the labour process* (London: Macmillan).
THOMPSON, Paul with WAILEY, Tony and LUMMIS, Trevor (1983) *Living the Fishing* (London: Routledge & Kegan Paul).
THORNE, Barrie and YALOM, Marilyn (eds) (1982) *Rethinking the Family* (London: Longmans).
TILLY, L.A. and SCOTT, J.W. (1978), *Women, Work and Family* (New York: Holt, Rinehart & Winston).
TUNSTALL, J. (1962) *The Fishermen* (London: McGibbon & Kee).

VOYDANOFF, P. (1980) 'Work Roles as Stressors in Corporate Families', *Family Relations*, 29.

WALBY, S. (1983) 'Patriarchal Structures: the case of unemployment', in E. Gamarnikow *et al.* (eds), *Gender Class and Work* (London: Heinemann).

WALBY, S. (1984) 'Approaches to the Study of Gender Relations in Employment and Unemployment', in B. Roberts *et al.* (eds), *New Approaches to Economic Life* (Manchester: Manchester University Press).

WALLMAN, Sandra (ed.) (1979) *Social Anthropology of Work*, ASA Monograph 19 (New York: Academic Press).

WILKINSON, F. (ed.) (1981) *The Dynamics of Labour Market Segmentation* (London: Academic Press).

ZUCKERMAN, M. and LUBIN, B. (1965) *Manual for the Multiple Affect Adjective Check List* (San Diego: Education and Industrial Testing Service).

Index